GREEN SYNTHESES

Volume 1

Green Syntheses Series

Series Editors
Pietro Tundo & John Andraos

Green Syntheses, Volume 1
Pietro Tundo and John Andraos

GREEN SYNTHESES

Volume 1

Edited by
Pietro Tundo
John Andraos

CRC Press
Taylor & Francis Group
Boca Raton London New York

CRC Press is an imprint of the
Taylor & Francis Group, an **informa** business

CRC Press
Taylor & Francis Group
6000 Broken Sound Parkway NW, Suite 300
Boca Raton, FL 33487-2742

First issued in paperback 2019

ISBN-13: 978-1-4665-1320-4 (hbk)
ISBN-13: 978-0-367-37863-9 (pbk)

Library of Congress Cataloging-in-Publication Data

Green syntheses / edited by Pietro Tundo, John Andraos.
2 volumes ; cm. -- (Green syntheses series)
Summary: "This book, the first volume in the new Proven Green Syntheses series, presents new reaction pathways and organic and inorganic catalysts involving and including fundamental chemistry. A unique feature of this new series, not found elsewhere, is to include green metrics as a key component of submissions of original works in order to substantiate the level of "greenness" of new chemical processes. It helps fill an expressed need and desire by both academia and industry to incorporate metrics analysis as a means to rigorously define efficiency and sustainability of chemical synthesis. "-- Provided by publisher.
Includes bibliographical references and indexes.
ISBN 978-1-4665-1320-4 (v. 1 : hardback)
1. Green chemistry. I. Tundo, Pietro, 1945- editor of compilation. II. Andraos, John, editor of compilation.

TP155.2.E58G77 2014
660--dc 3   2014007632

Visit the Taylor & Francis Web site at
http://www.taylorandfrancis.com

and the CRC Press Web site at
http://www.crcpress.com

Contents

Preface

The importance of reproducibility of results is increasing together with the accuracy for experimental detail. Moreover, many new methods and protocols directed to the syntheses of useful intermediates have been and are being developed to specifically address green and sustainable chemistry principles. Therefore, an appreciation of ethical conduct and efforts to uphold such behavior are needed to maintain our credibility in the eyes of the public. This is particularly relevant to green chemistry, since as scientists we have an important responsibility to society and to the tradition of science.

Highlighting the importance of green metrics, the Green Syntheses Series focuses on how to reliably substantiate and validate the level of "greenness" of chemical processes because substantiation of claims made is important for the credibility of green chemistry as a serious scientific discipline. The Green Syntheses Series evaluates by rigorous metrics the greenness of chemical synthesis proposed. We determine appropriate material efficiency green metrics in order to compare syntheses provided by authors with those published in the past. This is a new concept in green chemistry—everybody wants to be green, but now we need to go forward with ethical intent by supplying appropriate proof. We should move forward from the generic and spontaneous intention to rigorous measures, from rough and casual estimation to precision.

The main purpose of launching the Green Syntheses Series is to reverse these shortcomings so that new smarter chemistry can be achieved and that these innovations can be verified rigorously by some means. We hope that the present work demonstrates what future publications might look like if green principles are followed and that also incorporates the important ethical aspect of supplying rigorous evidence of greenness of a given synthesis protocol using metrics analysis. Also, we are quite conscious that metrics, as a key sub-branch of green chemistry, is a rapidly evolving field and is subject to error and revision.

Our goal should be that as chemists we are able to work in an open atmosphere of trust, in which we can apply ourselves meticulously and with enthusiasm, so that we can continue to contribute responsibly to the solutions of problems facing humanity in these rapidly evolving and challenging times.

Pietro Tundo
Venice, Italy

John Andraos
Toronto, Canada

About the Editors

Pietro Tundo is professor of organic chemistry at Ca' Foscari University of Venice, Italy. He has been a guest researcher and teacher at College Station (Texas, 1979–1981), Potsdam (New York, 1989–1990), Syracuse (New York, 1991–1992), and Chapel Hill (North Carolina, 1995).

Dr. Tundo's scientific interests are in the field of organic synthesis in selective methylations with low environmental impact, continuous flow chemistry, chemical detoxification of contaminants, hydrodehalogenation under multiphase conditions, phase-transfer catalysis (gas–liquid phase-transfer catalysis (GL-PTC)), synthesis of crown-ethers and functionalized cryptands, supramolecular chemistry, and finally, heteropolyacids.

He is holder of the UNESCO Chair on Green Chemistry (UNTWIN No. 731) and author of approximately 260 scientific publications and 30 patents. He authored the book *Continuous Flow Methods in Organic Synthesis* (E. Horwood, 1991) and is editor of approximately 15 books.

Dr. Tundo was president of the Organic and Biomolecular Chemistry Division of the International Union of Pure and Applied Chemistry (IUPAC) (biennium 2007–2009) and presently is chairman of the Working Party on Green and Sustainable Chemistry of the European Association for Chemical and Molecular Sciences (EuCheMS). He is founder of the IUPAC International Conferences Series on Green Chemistry, and was awarded the Kendall Award, with Janos Fendler, by the American Chemical Society in 1983 and An Intelligent Future by the Federchimica (Italian Association of Chemical Industries) in 1997.

Dr. Tundo has coordinated many institutional and industrial research projects (EU, NATO, Dow, ICI, Roquette) and was director of the 10 editions of the annual Summer School on Green Chemistry (Venezia), the latter sponsored by the EU, UNESCO, and NATO.

John Andraos earned a PhD in physical organic chemistry in 1992 from the University of Toronto. During his appointment as lecturer and course director at York University (1999–2009), he launched the first industrial and green chemistry course in the history of the Department of Chemistry. The Green Chemistry Education Roundtable at the U.S. National Academy of Sciences recognized this course as the most innovative green chemistry course from Canada in 2005. He has worked on consulting assignments with the pharmaceutical industry, including Merck, Bioverdant, Scientific Update, Apotex Pharmaceuticals, and Row2 Technologies, on various aspects of synthesis optimization and green chemistry education. Dr. Andraos's research spans several disciplines in organic chemistry that utilize mathematical analysis, including reaction kinetics, retrosynthesis and synthesis planning, and the creation of reaction and ring construction databases. He is the author of 50 scientific papers in refereed journals, 8 book chapters on the subject of green chemistry metrics and education, and the book *The Algebra of Organic Synthesis: Green Metrics,*

Design Strategy, Route Selection, and Optimization (CRC Press/Taylor & Francis, 2012). In 2000 he launched the CareerChem website (http://www.careerchem.com/ MainFrame.html), which is an in-depth resource for tracking and cataloging all named things in chemistry and physics, chronicling the development of chemistry through scientific genealogies, and supplying career information to young researchers and students for placement in academic and industrial positions worldwide. Dr. Andraos is now a full-time consultant with CareerChem in Toronto, Canada.

Contributors

Lo'ay A. Al-Momani
Department of Chemistry
Tafila Technical University
Tafila, Jordan

John Andraos
CareerChem
Toronto, Canada

Fabio Aricò
Dipartimento Scienze Ambientali,
   Informatica e Statistica
Università Ca' Foscari di Venezia
Venezia, Italy

Eleonora Ballerini
CEMIN—Dipartimento di Chimica
Biologia e Biotechnologie
Università di Perugia
Perugia, Italy

Arlene G. Corrêa
Departamento de Química
Universidade Federal de São Carlos
São Paulo, Brazil

Fidel Cunill
Faculty of Chemistry
University of Barcelona
Barcelona, Spain

Anna M. Deobald
Departamento de Química
Universidade Federal de São Carlos
São Paulo, Brazil

Amarajothi Dhakshinamoorthy
Centre for Green Chemistry Processes
and
School of Chemistry
Madurai Kamaraj University
Tamil Nadu, India

Matthew R. Dintzner
Department of Pharmaceutical and
   Administrative Sciences
Western New England University
Springfield, Massachusetts

Jordi Guilera
Faculty of Chemistry
University of Barcelona
Barcelona, Spain

Liang-Nian He
State Key Laboratory
and
Institute of Elemento-Organic Chemistry
Nankai University
Tianjin, People's Republic of China

Zhenshan Hou
Research Institute of Industrial Catalysis
East China University of Science and
   Technology
Shanghai, People's Republic of China

Montserrat Iborra
Faculty of Chemistry
University of Barcelona
Barcelona, Spain

Ermal Ismalaj
CEMIN—Dipartimento di Chimica
Biologia e Biotechnologie
Università di Perugia
Perugia, Italy

Kuppusamy Kanagaraj
School of Chemistry
Madurai Kamaraj University
Tamil Nadu, India

Márcio W. Paixão
Departamento de Química
Universidade Federal de São Carlos
São Paulo, Brazil

Kasi Pitchumani
Centre for Green Chemistry Processes
and
School of Chemistry
Madurai Kamaraj University
Tamil Nadu, India

Ferdinando Pizzo
CEMIN—Dipartimento di Chimica
Biologia e Biotechnologie
Università di Perugia
Perugia, Italy

Yunxiang Qiao
Research Institute of Industrial Catalysis
East China University of Science and
    Technology
Shanghai, People's Republic of China

Eliana Ramírez
Faculty of Chemistry
University of Barcelona
Barcelona, Spain

Daniel G. Rivera
Faculty of Chemistry
University of Havana
Havana, Cuba

Gustavo P. Romanelli
Centro de Investigación y
    Desarrollo en Ciencias
    Aplicadas "Dr. Jorge J. Ronco"
    (CINDECA-CCT-CONICET)
and
Facultad de Ciencias Agrarias y
    Forestales
Universidad Nacional de La Plata
La Plata, Argentina

Diego M. Ruiz
Facultad de Ciencias Agrarias y
    Forestales
Universidad Nacional de La Plata
La Plata, Argentina

Ángel G. Sathicq
Centro de Investigación y
    Desarrollo en Ciencias
    Aplicadas "Dr. Jorge J. Ronco"
    (CINDECA-CCT-CONICET)
Universidad Nacional de La Plata
La Plata, Argentina

Qing-Wen Song
State Key Laboratory
and
Institute of Elemento-Organic Chemistry
Nankai University
Tianjin, People's Republic of China

Javier Tejero
Faculty of Chemistry
University of Barcelona
Barcelona, Spain

Pietro Tundo
Dipartimento Scienze Ambientali,
    Informatica e Statistica
Università Ca' Foscari di Venezia
Venezia, Italy

Luigi Vaccaro
CEMIN—Dipartimento di Chimica
Biologia e Biotechnologie
Università di Perugia
Perugia, Italy

Lucas C. C. Vieira
Departamento de Química
Universidade Federal de São Carlos
São Paulo, Brazil

1 Introduction to the Green Syntheses Series

Pietro Tundo and John Andraos

CONTENTS

PHILOSOPHY AND PURPOSE

Despite the continued involvement of the industrial and scientific community in the field of sustainable chemistry, it was only between 1996 and 1997 that the term *green chemistry* was first used [1, 2]. The definitions of *green chemistry* and *sustainable chemistry* have been, and still are, the subject of a long ongoing debate. Herein the term *green chemistry* will be used according to the International Union of Pure and Applied Chemistry (IUPAC) definition that states: "Green Chemistry includes the invention, design, and application of chemical products and processes to reduce or to eliminate the use and generation of hazardous substances" [3].

In the late 1990s, with the new millennium looming, interest in green chemistry became widespread. In 1998 upon a U.S. Environmental Protection Agency (USEPA) proposal, the Organization for Economic Cooperation and Development (OECD) instituted a directive committee for the development of sustainable chemistry and finalized a program called Sustainable Chemistry that included chemistry aimed at pollution prevention and better industrial performance. Representatives from government, industry, and academic groups from 22 countries discussed policy and program aspects of implementing sustainable chemistry activity at the Venice Workshop (October 1998). These were subsequently approved at the OECD meeting in Paris (June 6, 1999) [4]. As a result of this meeting the following seven research areas in green/sustainable chemistry were identified:

1. *Use of alternative feedstocks*: The use of feedstocks that are renewable rather than depleting and less toxic to human health and the environment.
2. *Use of innocuous reagents*: The use of reagents that are inherently less hazardous and are catalytic whenever feasible.
3. *Employing natural processes*: Use of biosynthesis, biocatalysis, and biotech-based chemical transformations for efficiency and selectivity.

1

4. *Use of alternative solvents*: The design and utilization of solvents that have reduced potential for detriment to the environment and serve as alternatives to currently used volatile organic solvents, chlorinated solvents, and solvents that damage the natural environment.

5. *Design of safer chemicals*: Use of molecular structure design—and consideration of the principles of toxicity and mechanism of action—to minimize the intrinsic toxicity of the product while maintaining its efficiency of function.

6. *Developing alternative reaction conditions*: The design of reaction conditions that increase the selectivity of the product and allow for dematerialization of the product separation process.

7. *Minimizing energy consumption*: The design of chemical transformations that reduce the required energy input in terms of both mechanical and thermal inputs and the associated environmental impacts of excessive energy usage.

In many countries, the word *green* is associated with many and different subjects. In the United States, green symbolizes the color of money. The U.S. dollar bill is nicknamed the greenback. Therefore, in markets that use the U.S. dollar as currency, green carries a connotation of money, wealth, and commerce. Green is considered the traditional color of Islam because of its association with nature, as shown by the national flag of Saudi Arabia. Green is a symbol of Ireland, which is often referred to as the Emerald Isle (green also represents St. Patrick's Day). In Dante's *Divine Comedy*, green is the color used to symbolize hope, as well as in the Roman Catholic Church, where green is a traditional color symbolizing hope and the tree of life. The color green is often used as a symbol of sickness. In Japan, green indicates safety and luxury. In the Russian language green is synonymous with "not ripe." In the English language the word *green* used as an adjective has several meanings: (1) the color green, (2) covered by green growth of foliage, (3) youthful and vigorous, (4) pleasantly alluring, (5) not ripened or mature, (6) fresh and new, (7) marked by a sickly and nauseating appearance, (8) affected by intense envious emotion, (9) unseasoned or freshly sawed wood, (10) deficient in training, knowledge, or experience, (11) not fully qualified or experienced in a particular function, and (12) indicating that everything is in order and ready to process according to plan [5]. In our present context, since science, and in particular chemistry, has proved to be valuable and successful in solving many world problems (human diseases, efficient transport, improved agricultural yields, water purification, human welfare, etc.), everyone is seeking solutions to their own problems. But because different nations have different needs, confusion often arises about the meaning that each group of people attributes to this word, since all of them think about chemistry through green chemistry. Thus, each country requires green chemistry to be involved in solving specific questions: Latin America to the exploitation of renewable resources, the Arab regions to water quality and treatment, and the Far East countries to anti- and de-pollution issues.

Adoption of green chemistry principles is now possible because presently developed technologies have the capacity to build new protocols for manufacturing molecular species, and hence value-added industrial products that modern civilization needs to enjoy a high standard of living. This scenario is actually not only new, but also challenging, since it opens a direct dialog with social sciences and humanities sciences. Hence, implementing green practices is a multidisciplinary and multidimensional endeavor that necessitates acquiring knowledge and skill in various subjects and dealing with people of various backgrounds. Actually, this is a great challenge and opportunity for chemists moving from being perceived as polluters of the planet to people who are able to carry out and address a dialog on sustainable development together with politicians, economists, entrepreneurs, philosophers, and other professionals. A distinctive and emblematic example of this change is the recent 2013 Nobel Prize for Peace awarded to the Organization for the Prohibition of Chemical Weapons (OPCW).

However, we chemists have to be more and more conscious of our ethical duties, as we will be subject to many demands and requests from society—great expectations by humankind means that green chemistry will face other questions, as we will be requested to be more precise and secure. Achieving public recognition that chemistry is a crucial part of the solutions to living within safe operating spaces and environments for human development and survival is therefore crucial and critical for the reputation of chemistry as a science in the service of humanity. As such, sustainability is a profound challenge for science and technology that chemists are well positioned to address—from an understanding of the molecular basis of the natural and human environment to the developments of new products and energy sources on which a sustainable future will depend. We are evolving from the study and mitigation of undesired effects to inventing new green technologies; from legislating environmental protection and clean energy, carrying out complex life cycle analyses, and monitoring climate change to developing inherently safe processes and products through new reaction pathways for low CO_2 emissions with the aim of achieving carbon-neutral processes. Two example strategic approaches are possible that illustrate new kinds of thinking:

- The bottom-up approach deals with new reactions and processes being investigated on a lab scale in academies, according to OECD recommendations, that could be proposed for exploitation by industry through scaling up for future production (the most appropriate term would be *green chemistry*).
- The top-down approach moves from the industrial needs in basic and applied research; it is directed to design alternatives to harmful chemicals for industrial products and polluting processes (the most appropriate term would be *sustainable chemistry*).

Since the birth of industrial chemistry in the late nineteenth century with the advent of the dyestuffs industry in Germany and England, chemists have invested enormous effort into designing chemicals with various applications and properties, ranging from medicines and cosmetics to modern state-of-the-art materials and molecular machines. However, for the most part, their work demonstrated a quite surprising lack of interest in taking hazards into consideration in the design process. In 2013 the third issue of *Accounts of Chemical Research* was dedicated to presenting the latest review articles on toxicity and environmental issues associated with nanomaterials, areas that were long neglected in research. The goal of designing safer chemicals can be achieved through several different strategies (i.e., metrics and computational studies), the choice of which is largely dependent on the amount and quality of information that exists on the particular substance and synthetic pathways on how to synthesize it. The control of chemicals in the environment is already studied and managed by many other disciplines, agencies, and countries that are moving in this direction:

1. The EU has recently established Registration, Evaluation, Authorization and Restriction of Chemicals (REACH), a single integrated system for the registration, evaluation, and authorization of chemicals, together with its European Chemicals Agency (ECHA). Today the European and international legislation and EU directives for environmental protection are becoming stricter and recognize that there is a growing need for replacement of harmful compounds at a productive and end-user level.
2. The USEPA Presidential Green Chemistry Challenge Awards, established in 1996, promote the environmental and economic benefits of developing and using novel green chemistry. These annual awards recognize outstanding chemical technologies that incorporate the principles of green chemistry into chemical design, manufacture, and use in academic, industrial, and business sectors.
3. The United Nations Framework Convention on Climate Change (UNFCCC), the international environmental treaty, whose mission is to stabilize greenhouse gas concentrations in the atmosphere at a level that would prevent dangerous anthropogenic interference with the climate system, recognizes that developed countries are principally responsible for the current high levels of greenhouse gas emissions in the atmosphere as a result of more than 150 years of industrial activity.

But, the substitution of harmful compounds means that we will need to devote and supply more intelligence and attention, together with a rigid control of the results. The kind of innovative chemistry demanded is necessarily considerably more difficult to achieve and illustrates why it has not been adopted before. How should we move in the context of chemistry in general and of green chemistry in particular? We have already adopted green chemistry principles; however, we are asked now to be more precise and circumspect about claims of achieving reaction or plan "greenness." In his *Le Discours de la Méthode* Descartes observed, "Le bon sens est la chose du monde la mieux partagée, car chacun pense en être bien pourvu" (Common sense is the most widely shared commodity in the world, for every man is convinced that he is well supplied with it) [6].

However, in recent years the incidence of scientific misconduct has increased, such as duplicate publications, plagiarism, retractions, scientific fraud, scientific misconduct, and self-plagiarism [7–11]. Calls for a reevaluation of the peer-review process have also appeared [12]. We should endorse responsibilities in order to maintain the integrity of chemistry within the scientific community and restore the confidence of the general public in chemistry as a responsible contributor to the solutions of the global problems facing mankind in this century and beyond. This will be particularly relevant with respect to green chemistry where an ethical requirement is associated with scientific rigor. Although green synthesis does not face the previously mentioned kinds of fraud, the sensitivity to be rigorous will be more and more important in the future. Despite the fact that overall trust in science remains generally solid, the scientific community risks losing its credibility among the general public if this point is not seriously addressed. Therefore, an appreciation for the significance of ethical conduct and efforts to maintain it are needed to preserve our credibility in the eyes of the public.

Accurate documentation of the true content of a synthetic protocol is not a trivial matter. The procedures used by academic researchers often fall far short of accurate determination of true reaction material and energy efficiency. Omission of amounts of workup and purification materials used is routine in journal publications. The order of addition of reagents, controls on temperature and pressure, etc., reaction time, and how the reaction was worked up and the product isolated are important matters to deal with if the given protocol has any hope of being reproduced as originally stated. The inability to reproduce procedures in the chemistry literature not only casts doubt on the reported chemistry, but also erodes the reputation of chemistry as a proper scientific discipline since the characteristic of reproducible experimental outcomes is a fundamental defining feature of a subject to be called a science. These shortcomings have been highlighted and discussed in the past [13–17], and recently prominent chemists and editors of journals have begun to change editorial policies to address them, especially in the context of ethics demanded by green chemistry in its application to process chemistry, the pharmaceutical industry, and new technologies [18–29]. Given this demanding scenario, challenges and responsibilities are in front of us chemists, and this new series responds to and faces the question of being secure in organic and inorganic syntheses.

The aim of *Green Syntheses* is to evaluate by rigorous metrics the greenness of chemical syntheses proposed, which has not been covered so far. In this work, the first of its kind, we determine appropriate material efficiency green metrics, as outlined in the next section, in order to compare syntheses provided by authors with those already published in the past. This is a new concept in green chemistry—everybody wants to be green, but we need now to go forward with ethical intent by supplying appropriate proof.

As Descartes suggested, we should try to move forward from the generic and spontaneous intention to rigorous measures, from rough and casual estimation to precision. A point that needs to be made clear is that the present work is not to be viewed as casting blame on past practices of designing, doing, and ultimately reporting chemistry protocols in the literature, whether they were reported as green or not. Instead, what is hoped is that the present work demonstrates what future publications might look like if green principles are followed and also incorporate the important ethical aspect of supplying rigorous evidence of greenness of a given synthesis protocol using metrics analysis. Also, we are well conscious that metrics, as a key subbranch of green chemistry, is a rapidly evolving field and is subject to error and revision. With this in mind, we face this risk and undertake the publication of this first volume, watchfully taking care of criticism and suggestions for further refinement.

METRICS USED

From the 12 principles of green chemistry [30] listed in Figure 1.1, which are now well accepted, we can highlight those that may be parameterized by metrics that can measure the degree of greenness adhered to by those principles. Essentially, Figure 1.1 shows explicitly which of these commonsense statements are amenable to precise quantification by some kind of measure so that true, verifiable, and fair claims of greenness can be made. Principles 1, 2, 8, and 9 are connected to material consumption, principle 6 is connected to energy consumption, and principles 3, 4, and 5 are connected to life cycle assessment. This last category covers both occupational health and safety of workers manufacturing chemical products and consumers using them, and environmental impact of synthesized chemicals when they are released into the biosphere, which is composed of four components: air, water, soil, and sediment. A complete and fair metrics analysis for a given reaction or synthesis plan should ideally cover all three major classifications; however, there are a number of challenges to accomplish this. First, the question of cost and time investment may be prohibitive or demanding. Second, there

1. **Prevention:** It is better to prevent waste than to treat or clean up waste after it has been created.
2. **Atom economy:** Synthetic methods should be designed to maximize the incorporation of all materials used in the process into the final product.
3. **Less hazardous chemical syntheses:** Wherever practicable, synthetic methods should be designed to use and generate substances that possess little or no toxicity to human health and the environment.
4. **Designing safer chemicals:** Chemical products should be designed to effect their desired function while minimizing their toxicity.
5. **Safer solvents and auxiliaries:** The use of auxiliary substances (solvents, separation agents, etc.) should be made unnecessary wherever possible and innocuous when used.
6. **Design for energy efficiency:** Energy requirements of chemical processes should be recognized for their environmental and economic impacts and should be minimized. If possible, synthetic methods should be conducted at ambient temperature and pressure.
7. **Use of renewable feedstocks:** A raw material or feedstock should be renewable rather than depleted whenever technically and economically practicable.
8. **Reduce derivatives:** Unnecessary derivatization (use of blocking groups, protection/ deprotection, temporary modification of physical/chemical processes) should be minimized or avoided if possible, because such steps require additional reagents and can generate waste.
9. **Catalysis:** Catalytic reagents (as selective as possible) are superior to stoichiometric reagents.
10. **Design for degradation:** Chemical products should be designed so that at the end of their function they break down into innocuous degradation products and do not persist in the environment.
11. **Real-time analysis for pollution prevention:** Analytical methodologies need to be further developed to allow for real-time, in-process monitoring, and control prior to the formation of hazardous substances.
12. **Inherently safer chemistry for accident prevention:** Substances and the form of a substance used in a chemical process should be chosen to minimize the potential for chemical accidents, including releases, explosions, and fires.

FIGURE 1.1 Twelve principles of green chemistry as outlined by Anastas and Warner. Bold principles are amenable to metrics parameterization. (From Anastas, P. T., and Warner, J. C., *Green Chemistry: Theory and Practice*, Oxford University Press, New York, 1998 [30].)

is the pressing problem of dealing with unavailable and unreliable data to carry out a thorough and balanced evaluation. This is particularly relevant to life cycle assessment and has been discussed thoroughly elsewhere [31]. Third, and most importantly, the central inertia is a cultural and educational one in the community of chemists rather than a scientific one. As a standard part of their accreditation and certification as chemistry professionals, synthetic chemists are not trained in the necessary skills to accomplish such a task, which include balancing chemical equations, and a combination of knowledge in toxicology, ecology, thermodynamics, engineering, and other physical sciences. Traditionally, their main focus is on successfully synthesizing, identifying, and verifying product structures of interest while ignoring the identity and fate of all other side products and by-products. They concern themselves most with obtaining their desired target product from a chemical reaction with the correct structural assignments, but less so with the impacts of that reaction outside the reaction vessel.

Since its inception in the early 1990s as a philosophy based on 12 general guiding principles, the newly coined field of green chemistry has enjoyed a honeymoon phase where claims of greenness were justified based on one or two criteria. The main drive to reduce overall waste focused on solvent demand because it was obvious that these materials represent the largest fraction of bulk material inputted into a reaction vessel. Hence, justification of greenness in a chemical process involved criteria that were based on either using reduced amounts of solvents, choosing alternative safer and environmentally friendly solvents, or using no solvents at all in so-called solventless reactions. Another well-adopted strategy was to employ a two-pronged approach where the solvent problem was dealt with by any one of the above means, and in addition, the innovation was coupled to the introduction of a new green technology such as microwave heating, mechanochemical activation, continuous flow technology, ultrasound activation, or a recycling technology. In most reports claims were not rigorously checked for either material consumption or energy consumption, and so pronouncements of greenness, though made assertively, were in fact tenuous.

The main purpose of launching the Green Syntheses Series is to reverse these shortcomings so that new smarter chemistry can be achieved, and that these innovations can be verified rigorously by some means. Following Descartes' observation, this new series provides the much needed transition between making 12 general commonsense statements about the philosophy of green chemistry to the actualization and verification of it in laboratory practice. Substantiation of claims made is important for the credibility of green chemistry as a serious scientific discipline. In this vein, this inaugural volume has addressed primarily the issue of material efficiency metrics, which measure the amount of waste produced relative to desired product, or alternatively, the fraction of input materials that end up in the desired product. Both perspectives are complementary and describe the same concept from two points of view akin to viewing a glass of water as being both half full and half empty. The choice of materials metrics is evident because it is conceptually and practically the easiest category to deal with and is accessible to any practicing chemist. It is hoped that future volumes in this series will address the other categories of energy consumption and life cycle assessment as the wider community of chemists continues to adopt green chemistry ideas and practice in their work. Success in addressing these other metrics areas will be directly linked to the degree of disclosure of necessary parameters, such as energy consumption in experimental procedures, so it becomes a matter of standard protocol. The reports of Clark [32] and Leadbeater [33] are important milestones toward that end.

Throughout this volume the material metrics chosen were as follows: atom economy [34], E-factor [35], process mass intensity (PMI) [36], reaction yield, and the three E-factor constituents, E-kernel, E-excess, and E-aux, corresponding to waste materials arising from reaction by-products and unreacted reagents, excess reagents, and auxiliary materials from workup and purification procedures. The next section illustrates how these metrics may be determined for a given chemical reaction along with their explicit definitions. As part of the process of submissions to this series, contributors were required to present a single reaction or a two-step sequence to a known product. In each contributing example, full experimental details are given showing all quantities of materials used in the procedure. A thorough literature search was done on all past procedures for that same target molecule. Each reported synthesis was then evaluated according to the material efficiency metrics described above. The results are tabulated and ranked according to PMI. Authors discuss the green merits of their protocols in conjunction with the results of a thorough metrics analysis. This approach allows for a deeper and meaningful discussion of insights about synthesis strategy and performance characteristics of the new and prior cited plans. As appropriate, authors suggest future improvements to their protocols in light of the metrics analyses.

TUTORIAL EXAMPLE OF MATERIAL EFFICIENCY METRICS CALCULATIONS

In order to clarify the definitions and implementation of green metrics used in the Green Syntheses Series, a fully-worked-out illustrative example is given below. The reaction is a Suzuki cross-coupling taken from an *Organic Syntheses* experimental procedure [37]. Scheme 1.1 shows the fully balanced chemical equation for the transformation, and Table 1.1 summarizes the itemization of all input materials required along with their quantities. Since all required metrics are determined on the basis of mass units, all quantities of input materials are calculated in terms of mass in grams. In this protocol p-bromobenzaldehyde is the limiting reagent and 42.5 g of p-phenylbenzaldehyde product is collected after workup and purification.

From Scheme 1.1 the atom economy for the transformation is given by

$$AE = \frac{MW_{product}}{\sum MW_{reagents}} = \frac{182}{122 + 184.9 + 106 + 18} = 0.422 \tag{1.1}$$

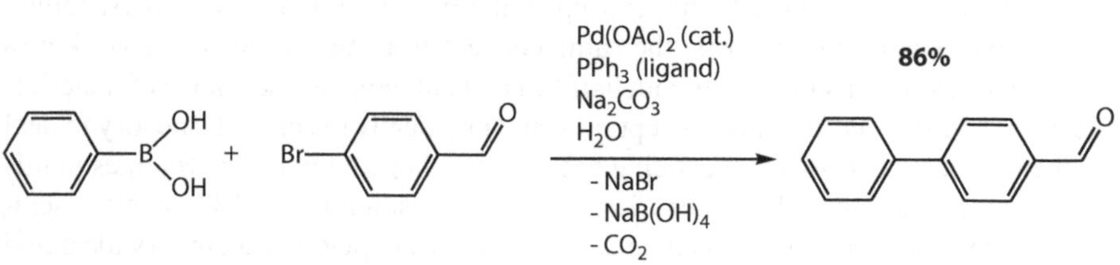

SCHEME 1.1   Example Suzuki cross-coupling reaction.

Hence, the percent atom economy is 42.2%. Since 42.5 g of product is collected, this translates to $42.5/182 = 0.234$ mole. When this molar amount is compared to that of the limiting reagent, the percent yield of the reaction is given by

$$\%Y = 100\left(\frac{\text{moles}_{\text{product}}}{\text{moles}_{\text{limiting-reagent}}}\right)\left(\frac{\text{stoich.coeff.}_{\text{limiting-reagent}}}{\text{stoich.coeff.}_{\text{product}}}\right)$$

$$= 100\left(\frac{0.234}{0.270}\right)\left(\frac{1}{1}\right) = 86.4 \tag{1.2}$$

TABLE 1.1
Summary of Quantities of Input Materials Used in Suzuki Cross-Coupling Reaction Shown in Scheme 1.1

Input Material	Molecular Weight (g/mol)	Density (g/mL)	Volume (mL)	Number of Moles	Mass (g)
Reactants					
p-Bromo-benzaldehyde	184.9			0.270	50
Phenylboronic acid	122			0.284	34.6
Na_2CO_3 (2 M aqueous solution)	106		162	0.324	34.3
Water	18	1	95	5.278	95
Catalysts/Ligands					
$Pd(OAc)_2$	224.4			0.00081	0.18
Triphenylphosphine	262			0.00244	0.64
Reaction Solvents					
n-Propanol		0.894	485		433.6
Water from 2 M Na_2CO_3 solution		1.2	162		160.1
Workup Materials					
Water		1	350		350
EtOAc		0.901	1000		901
5 wt% $NaHCO_3$ solution		1.0354	250		258.9
Saturated NaCl solution		1.1804	500		590.2
Darco G6					25
Na_2SO_4					50
Purification Materials					
Florisil					50
Celite					50[a]
EtOAc		0.901	200		180.2
Hexane		0.659	269		177.3
Methanol		0.792	47.3		37.5

[a] Assumed mass based on amount of florisil declared.

Summing all the masses of input materials, we obtain 3478.4 g. Therefore, the amount of waste produced in this transformation is 3478.4 − 42.5 = 3435.9 g. The overall E-factor and the PMI are given by Equations 1.3 and 1.4, respectively.

$$E_{total} = \frac{\sum mass_{waste}}{mass_{product}} = \frac{3435.9}{42.5} = 80.8 \tag{1.3}$$

$$PMI = \frac{\sum mass_{inputs}}{mass_{product}} = \frac{3478.4}{42.5} = 81.8 \tag{1.4}$$

We can see from Equations 1.3 and 1.4 that the connecting relation $PMI = 1 + E$ is verified. From Table 1.1 we determine that the sum of masses of auxiliary materials used in workup and purification is 2175.1 + 494.9 = 2670.0 g. Therefore, the E-factor contribution, E-aux, is given by

$$E_{aux} = \frac{\sum mass_{auxiliaries}}{mass_{product}} = \frac{2670}{42.5} = 76.8 \tag{1.5}$$

Since 0.234 mole of target product is collected and sodium bromide (MW 102.9), carbon dioxide (MW 44), sodium hydroxide (MW 40), and boric acid (MW 62) are also produced as reaction by-products as a consequence of this, we can determine the by-product contribution to the E-factor as follows:

$$E_{byproducts} = \frac{\sum mass_{byproducts}}{mass_{product}} = \frac{0.234(102.9 + 44 + 40 + 62)}{42.5} = \frac{58.2}{42.5} = 1.4 \tag{1.6}$$

Since p-bromobenzaldehyde is the limiting reagent, the stoichiometric molar amount of each input reagent is 0.270. The contribution to the total E-factor from excess reagents is given by

$$E_{excess} = \frac{\sum mass_{excess-reagents}}{mass_{product}}$$

$$= \frac{(0.284 - 0.27)122 + (0.324 - 0.27)106 + (5.278 - 0.27)18}{42.5}$$

$$= \frac{97.6}{42.5}$$

$$= 2.3 \tag{1.7}$$

The first, second, and third terms in the numerator refer to the excess masses of phenylboronic acid, sodium carbonate, and water, respectively. If 0.234 mole of product was collected, this means that 0.234 mole each of p-bromobenzaldehyde, phenylboronic acid, sodium carbonate, and water reacted. Since 0.27 mole of p-bromobenzaldehyde limiting reagent was inputted, we can conclude that $0.27 - 0.234 = 0.036$ mole of that reagent did not react. From the balanced chemical equation in Scheme 1.1 under stoichiometric conditions we can conclude that 0.036 mole of each input reagent did not react. Therefore, the contribution to E-total from unreacted reagents is given by

$$E_{unreacted} = \frac{\sum mass_{unreacted-stoichiometric-reagents}}{mass_{product}}$$

$$= \frac{0.036(184.9 + 122 + 106 + 18)}{42.5} = \frac{15.5}{42.5} = 0.4 \tag{1.8}$$

and the E-kernel contribution is given by

$$E_{kernel} = E_{byproducts} + E_{unreacted} = 1.4 + 0.4 = 1.8 \tag{1.9}$$

From these E-factor contributions we may verify that the sum of all such contributions is equal to E-total as shown below.

$$E_{total} = E_{kernel} + E_{excess} + E_{aux} = 1.8 + 2.3 + 76.8 = 80.9 \tag{1.10}$$

Accounting for round-off errors, this result is consistent with that given in Equation 1.3.

REFERENCES

1. Amato, I. *Science* 1993, 259, 1538.
2. Anastas, P. *Chem. Eng. News* 2011, 89(26), 62. (The first international conference was held in Venice, September 28–October 1, 1997: Green Chemistry, Challenging Perspectives.)
3. Tundo, P., Anastas, P., Black, D.StC., Breen, J., Collins, T., Memoli, S., Miyamoto, J., Polyakoff, M., Tumas, W. *Pure Appl. Chem.* 2000, 72, 1207 (see p. 1210).
4. (a) OECD Environmental Directorate. Joint Meeting of the Chemical Committee and the Working Party on Chemical Pesticides and Biotechnology, Paris, June 6, 1999; (b) IUPAC Green Chemistry Directory. Green Chemistry Research Areas. http://www.incaweb.org/transit/iupacgcdir/gcresearchareas.htm (accessed October 31, 2013).
5. *Webster's New Collegiate Dictionary*. Thomas Allen & Son Ltd., Toronto, 1976, p. 504.
6. *Oxford Dictionary of Quotations*, 2nd ed. Oxford University Press, London, 1955, p. 172.
7. Noyori, R., Richmond, J.P. *Adv. Synth. Catal.* 2013, 355, 3.
8. *Nat. Chem.* 2011, 3, 337 (editorial).
9. (a) Fang, F.C., Steen, R.G., Casadevall, A. *Proc. Nat. Acad. Sci.* 2012, 109, 17028. (b) Smith, A.B., III. *Org. Lett.* 2013, 15, 2893.
10. Laird, T. *Org. Proc. Res. Dev.* 2013, 17, 317.
11. Van Noorden, R. *Nature* 2011, 478, 26.
12. Resnik, D.B. *Am. Sci.* 2011, 99, 24.
13. Cornforth, J. *Chem. Br.* 1975, 11, 432.
14. Cornforth, J.W. *Austr. J. Chem.* 1993, 46, 157.

15. (a) Wernerova, M., Hudlicky, T. *Synlett* 2010, 2701. (b) Carlson, R., Hudlicky, T. *Helv. Chim. Acta.* 2012, 95, 2052.
16. Kovac, P. *Carbohydrate Chemistry: Proven Synthetic Methods*, vol. 1. CRC Press, Boca Raton, FL, 2012, p. xix.
17. Danheiser, R.L. *Org. Synth.* 2011, 88, 1.
18. Laird, T. *Org. Proc. Res. Dev.* 2011, 15, 305.
19. Laird, T. *Org. Proc. Res. Dev.* 2011, 15, 729.
20. Laird, T. *Org. Proc. Res. Dev.* 2012, 16, 1.
21. Tucker, J.L. *Org. Process Res. Dev.* 2006, 10, 315.
22. Tucker, J.L. *Org. Process Res. Dev.* 2010, 14, 328.
23. Carey, J.S., Laffan, D., Thomson, C., Williams, M.T. *Org. Biomol. Chem.* 2006, 4, 2337.
24. Constable, D.J.C., Dunn, P.J., Hayler, J.D., Humphrey, G.R., Leazer, J.L., Linderman, R.J., Lorenz, K., Manley, J., Pearlman, B.A., Wells, A., Zaks, A., Zhang, T.Y. *Green Chem.* 2007, 9, 411.
25. Warhurst, M. *Green Chem.* 2002, 4, G20.
26. Diehlmann, A., Kreisel, G. *Green Chem.* 2002, 4, G15.
27. Watson, W.J.W. *Green Chem.* 2012, 14, 251.
28. Welton, T. *Green Chem.* 2011, 13, 225.
29. Bennett, G.D. Green chemistry as an expression of environmental ethics. In Sharma, S.K., and Mudhoo, A. (eds.), *Green Chemistry for Environmental Sustainability*. CRC Press, Boca Raton, FL, 2010, p. 116.
30. Anastas, P.T., Warner, J.C. *Green Chemistry: Theory and Practice.* Oxford University Press, New York, 1998.
31. (a) Andraos, J. *Org. Process Res. Dev.* 2012, 16, 1482. (b) Andraos, J. *Org. Process Res. Dev.* 2012, 17, 175.
32. Gronnow, M.J., White, R.J., Clark, J.H., Macquarrie, D.J. *Org. Process Res. Dev.* 2005, 9, 516. (Corrigendum: Gronnow, M.J., White, R.J., Clark, J.H., Macquarrie, D.J. *Org. Process Res. Dev.* 2007, 11, 293.)
33. Devine, W.G., Leadbeater, N.L. *ARKIVOC* 2011, 5, 127.
34. Trost, B.M. *Science* 1991, 254, 1471.
35. Sheldon, R.A. *ChemTech* 1994, 24(3), 38.
36. Jiménez-González, C., Ponder, C.S., Broxterman, Q.B., Manley, J.B. *Org. Process Res. Dev.* 2011, 15, 912.
37. Huff, B.E., Koenig, T.M., Mitchell, D., Staszak, M.A. *Org. Synth.* 1998, 75, 53.

2 Application of Material Efficiency Metrics to Assess Reaction Greenness: Illustrative Case Studies from *Organic Syntheses*

John Andraos

CONTENTS

INTRODUCTION

In this chapter to the Green Syntheses Series the reader is introduced to a number of example reactions taken from *Organic Syntheses*. Since reports appearing in this well-known and venerable series have their experimental procedures independently checked as a strict requirement for publication [1], it seems only fitting that they would highlight the best possible example target molecules made by synthetic methods to illustrate how green metrics could be used to augment discussions of their synthesis performances. In addition to the already demonstrated reproducibility of experimental results, the purpose here is to illustrate what is possible with respect to reporting experimental procedures rigorously for the purpose of analyzing reaction and plan greenness based on simple material efficiency metrics. With this in mind, the selected *Organic Syntheses* protocols for specific target molecules are analyzed and ranked with past literature synthesis reports to the same target molecules. Overall, this approach goes the extra mile to give a fuller analysis of synthesis performance that covers synthesis strategy with respect to how a given target product is made most efficiently while producing the least amount of waste in the process. An important consequence of this added approach is that it gives a focused direction for future optimizations for the intended target molecule that is guided by the resulting metrics analysis for known routes to it. Hence, each future iteration of optimization would have an accompanying set of metrics that confirms overall waste reduction with respect to prior plans and specifically in which area, such as reaction solvent usage, auxiliary material usage, or excess reagent consumption. Up to now the main and customary focus of synthetic organic chemistry has been to showcase the diversity of possible ways to make compounds using both classical and newly developed synthesis methodologies.

Classical methodologies draw upon the large pool of traditional named organic reactions [2], whereas the modern ones draw upon the newly discovered methods that explore newer combinations of target bonds that could be made for a given structure, as well as demonstrating the performances of newly discovered catalysts. Since Wöhler's first synthesis of urea in 1828 [3], synthetic chemists have labored to increase the toolbox of reactions for functional group transformations and skeletal building reactions. The most important and challenging structural motif to build is the ring. It is this feature that has always attracted the attention of chemists to showcase their prowess in developing ring construction reactions, particularly as the structural complexity of ring systems increases. The most spectacular reactions are those that are able to make complex rings in the least number of steps and with the most target bonds formed in a single step. Terms such as *efficient*, *high yield*, and *concise* have been used widely to describe successful outcomes in synthesis literature. However, these attributes are usually made assertively on the basis of a single criterion or meritorious property of such syntheses, such as overall yield or number of steps, often without providing some kind of rigorous proof that takes into account all materials used. Though these two common parameters are important in gauging synthesis efficiency, they do not tell the whole story. Moreover, authors often do not dig into the literature deep enough to investigate past published plans of a given target molecule, limiting themselves to only one or two past decades. Consequently, they may be surprised to learn that a plan reported in the 19th or early 20th century may still hold the record for being the truly most efficient, and therefore most green, according to modern material efficiency parlance. A nice example of this is the 1935 synthesis of physostigmine by Percy Julian [4, 8b]. The present set of examples shows how a chemist could actually provide the necessary evidence by using a set of simple green metrics applied to the new synthesis, as well as to prior reported syntheses for the same target structure. This idea forms the basis of this inaugural volume and is illustrated by the contributing authors. In doing so, these authors demonstrate how green chemistry can evolve from a general philosophy based on 12 principles and statements made by the Organization for Economic Cooperation and Development (OECD) (see Introduction) [5] to a bona fide rigorous higher-order multidisciplinary science. Having said that, none of these authors have the final word on the syntheses of their chosen target molecules because the quest for optimization is a job that is never done. As new discoveries are made daily, there will always be a better way to carry out a given reaction or a better strategy to make a complex framework. This explains why new methodologies are often tested on well-known classical target molecules that are revisited over and over again, since they form the proving ground for synthesis. These ongoing endeavors to discover new synthesis strategies may be complemented with metrics analyses in a meaningful way that showcases both their novelty and degree of greenness.

The choice of reaction examples taken from *Organic Syntheses* was made on the basis of whether authors used keywords associated with green chemistry terminology in the text of their papers, such as *atom economy, green chemistry, sustainable*, and *multicomponent reaction*. For each example, accompanying schemes and summary tables of metrics performances are given, followed by a brief discussion of the ranking results. The layout of schemes is as follows: they are arranged by synthesis strategy type; in each step molecular weights are shown directly below structures and percent yields are given directly above them, and catalysts, ligands, and other additives are indicated as appropriate above reaction arrows. All chemical equations are fully balanced, showing by-products and stoichiometric coefficients. In order to explicitly highlight synthesis strategy patterns, target-forming bonds in the product structure are shown as bolded bonds in the schemes. The tables summarize the following metrics: atom economy, percent yield, E-kernel, E-excess, E-aux, E-total, and process mass intensity (PMI). Atom economy refers to the molecular weight ratio of the target product to the sum of reactants corresponding to a balanced chemical equation. Reaction yield is the mass ratio of actual versus theoretical target product. E-total is the total E-factor corresponding to the mass ratio of waste produced to target product collected. This is further broken down into its three contributing components to waste material: E-kernel, referring to reaction by-products as a direct consequence of producing the intended target product; E-excess, referring to excess reagent consumption; and E-aux, referring to auxiliary materials used in the workup and purification phases of a chemical reaction. The PMI is the mass ratio of all input materials to target product and has been recently adopted as the go-to metric by the pharmaceutical industry [6]. All of these metrics are calculated directly from masses of all materials declared in experimental procedures under the condition that all materials used are destined for waste except for the intended target product. Hence, it is imperative that full disclosure is adhered to in order for accurate determination of these metrics and reliable analysis of their rankings. Procedures described in *Organic Syntheses* fulfill this criterion in all categories except for masses of drying agents used, such as sodium or magnesium sulfate, and filtration materials such as celite or florisil. Disclosure for these substances was not always consistent. However, procedures appearing in experimental sections of journal articles or associated supplementary materials had the least consistency of detail. This persistent problem has been discussed before [7] and threatens to jeopardize progress in adopting green chemistry practice. Therefore, when missing data were encountered, the following assumptions were made: volumes of workup extraction solvents were set equal to the volume of reaction solvent, and the mass of drying agent was set to 2 g if the volume of reaction solvent was less than 20 mL, or 10 g if the reaction volume was between 100 and 500 mL. Undisclosed masses of silica gel and volumes of eluents in chromatographic procedures could not be estimated reliably; therefore, E-aux, E-total, and PMI values for such procedures appear as lower limits in the tables; i.e., values appear with a greater than inequality sign (>). Plans with such designations should be interpreted as having an associated uncertainty with respect to rankings. In the summary tables all reported literature syntheses are ranked according to PMI in ascending order; that is, best plans appear at the top with the lowest PMI values. The performances of plans referring to checked reports appearing in *Organic Syntheses* are shown in italics in the tables. For convenience, references to reactions are shown directly below the tables as footnotes. If plans to a given target molecule involved more than one reaction, then all plans to that target were evaluated using the previously described algorithm [8] using 1 mole as the basis scale for all plans. That algorithm takes into account the appropriate scaling factors for all materials used in a multistep synthesis plan when reactions are linked together sequentially.

EXAMPLE 2.1: 1,3-DIPHENYL-2-(1-PHENYLETHYL)-PROPANE-1,3-DIONE

This compound has been made by four strategies that all begin using 1,3-diphenyl-propane-1,3-dione as a reagent and some other coupling partner, such as styrene via direct addition (Scheme 2.1), 1-phenylethanol via a condensation (Scheme 2.2), phenylethane via oxidative addition (Scheme 2.3), and (1-bromoethyl)-benzene and

Liu, P.N.; Zhou, Z.Y.; Lau, C.P. *Chem. Eur. J.* **2007**, 13, 8610

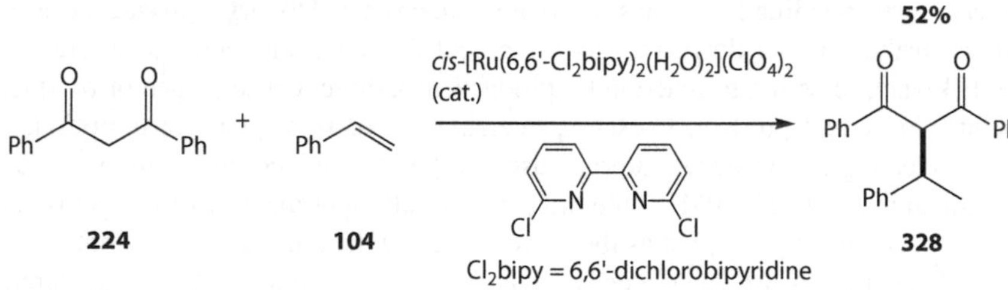

Das, B.; Krishnaiah, M.; Laxminarayana, K.; Damodar, K.; Kumar, D.N. *Chem. Lett.* **2009**, 38, 42

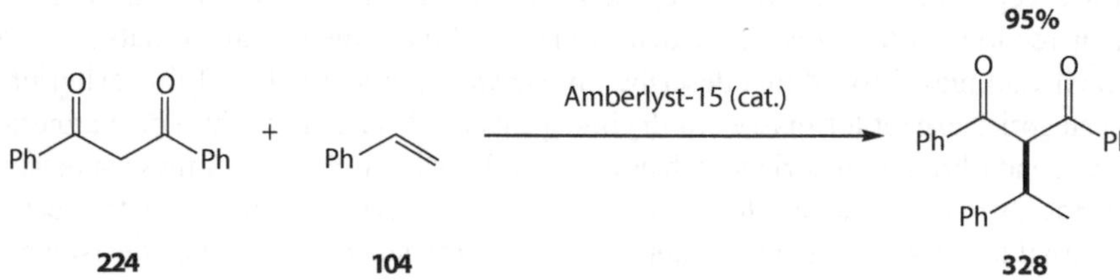

SCHEME 2.1 Synthesis plans to make 1,3-diphenyl-2-(1-phenylethyl)-propane-1,3-dione using styrene.

Yao, X.; Li, C.J. *J. Am. Chem. Soc.* **2004**, 126, 6884

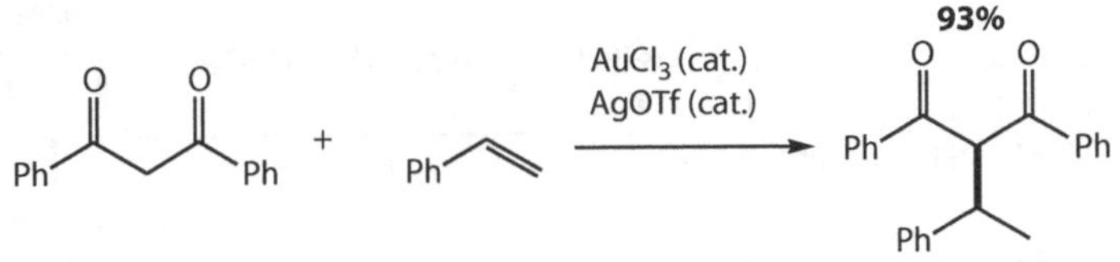

224 **104** **328**

Yao, X.; Li, C.J. *J. Org. Chem.* **2005**, 70, 5752

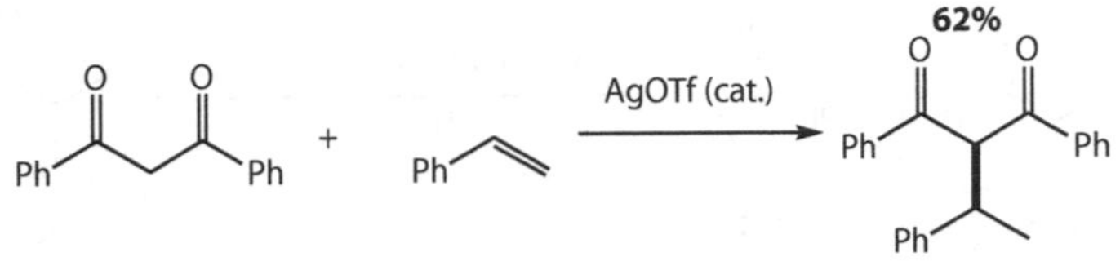

224 **104** **328**

Wang, G.W.; Shen, Y.B.; Wu, X.L.; Wang, L. *Tetrahedron Lett.* **2008**, 49, 5090

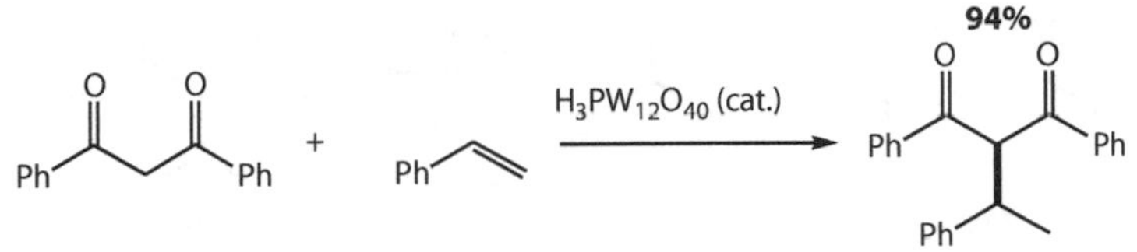

224 **104** **328**

Li, C.J.; Yao, X. *Org. Synth.* **2007**, 84, 222

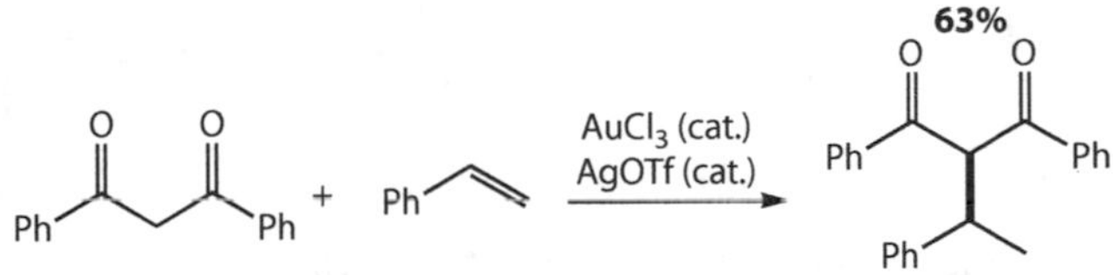

224 **104** **328**

SCHEME 2.1 ***(Continued)***

4-methyl-N-(1-phenyl-ethyl)-benzenesulfonamide via catalyzed substitution reactions (Scheme 2.4). From an atom economical point of view, the best strategy is to use styrene since no by-products are formed. The second best is to use 1-phenylethanol, which produces water as a by-product. If bromide and tosylamide are leaving groups instead of hydroxide, then the atom economy is reduced further compared to the condensation strategy. The overall worst strategy is oxidative addition, since an oxidizing agent is required that necessarily gets reduced, thus producing significant by-products. This is a nice tutorial example of carbon-carbon bond formation showcasing how atom economy alone can be used to elucidate efficient synthesis strategy. From the ranking results shown in Table 2.1, the best plan's ranking (Das, 2009) is unreliable since chromatographic materials were not disclosed. This was a problem throughout,

Liu, P.N.; Zhou, Z.Y.; Lau, C.P. *Chem. Eur. J.* **2007**, 13, 8610

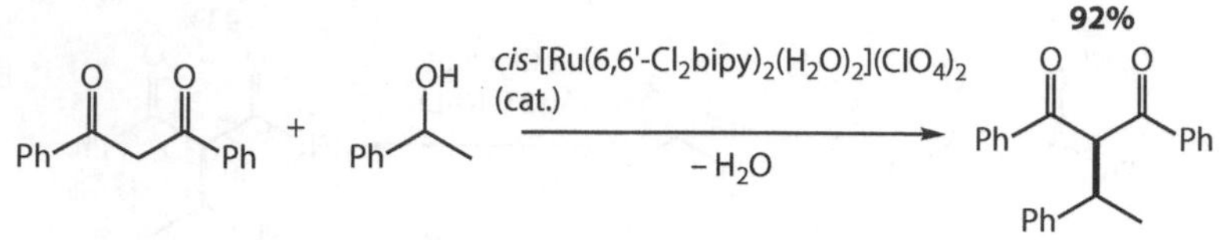

224 122 328

Liu, P.N.; Xia, F.; Wang, Q.W.; Ren, Y.J.; Chen, J.Q. *Green Chem.* **2010**, 12, 1049

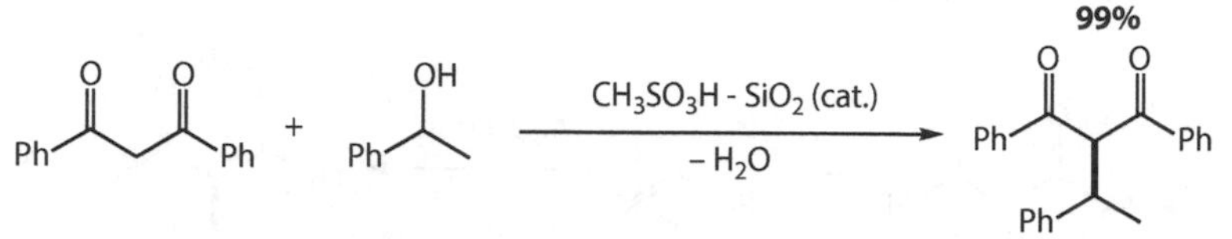

224 122 328

Li, Z.; Duan, Z.; Wang, H.; Tian, R.; Zhu, Q.; Wu, Y. *Synlett* **2008**, 2535

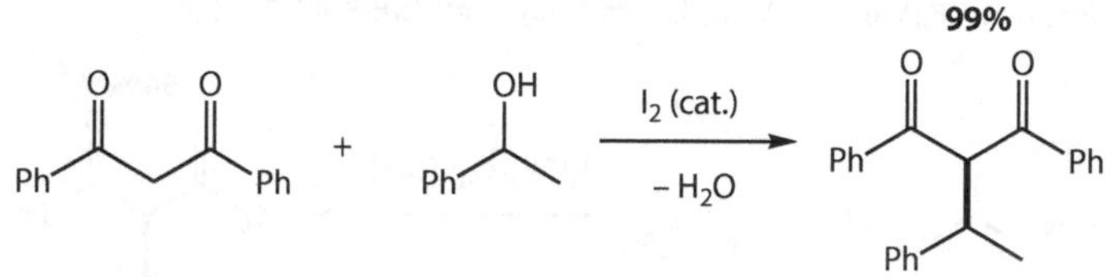

224 106 328

Qureshi, Z.S.; Deshmukh, K.M.; Tambade, P.J.; Bhanage, B.M. *Tetrahedron Lett.* **2010**, 51, 724

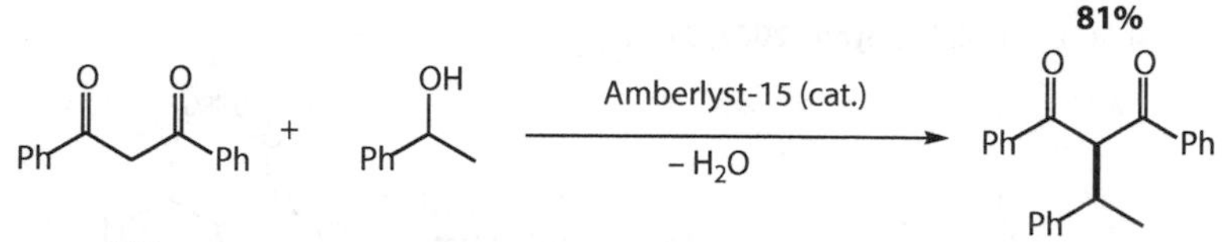

224 106 328

Xia, F.; Le Zhao, Z.; Liu, P.N. *Tetrahedron Lett.* **2012**, 53, 2828

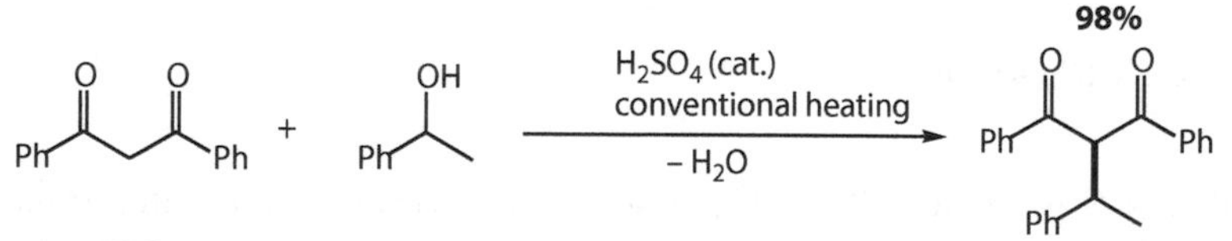

224 106 328

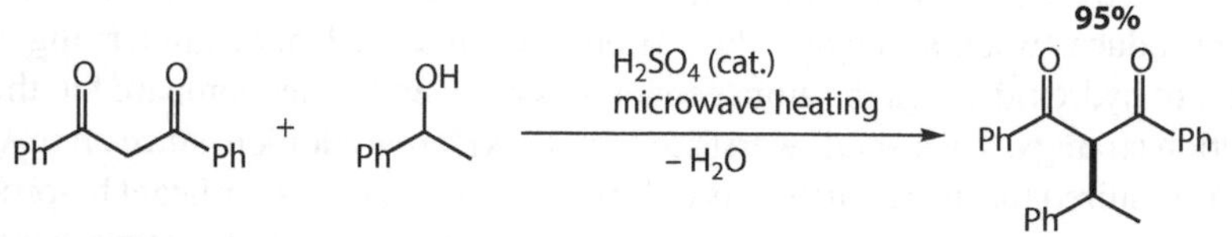

224 106 328

SCHEME 2.2 Synthesis plans to make 1,3-diphenyl-2-(1-phenylethyl)-propane-1,3-dione using 1-phenyl-ethanol.

Borduas, N.; Powell, D.A. *J. Org. Chem.* **2008**, 73, 7822

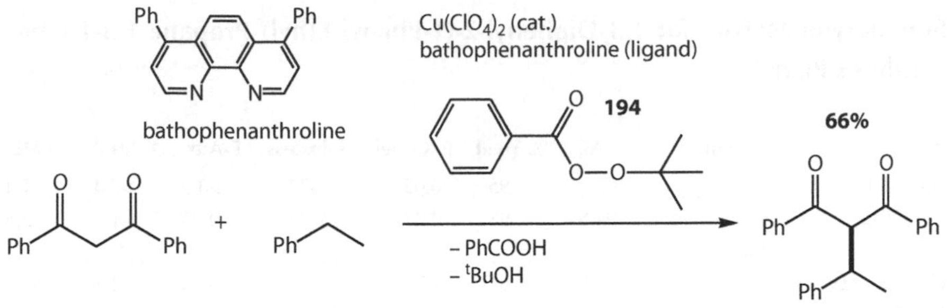

224 **106** **328**

Ramesh, D.; Ramulu, U.; Rajaram, S.; Prabhakar, P.; Venkateswarlu, Y. *Tetrahedron Lett.* **2010**, 51, 4898

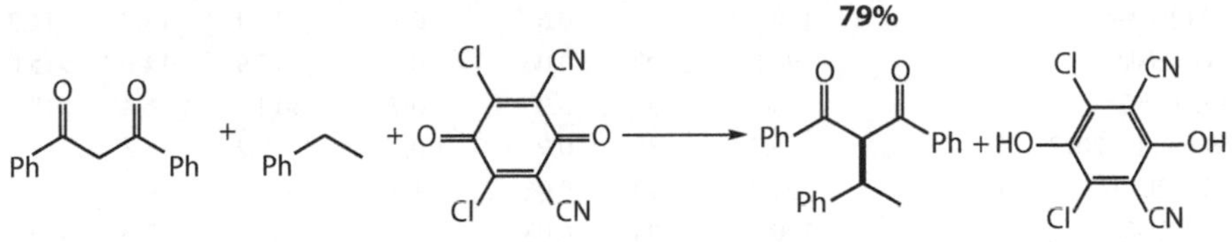

224 **106** **226.9** **328** **228.9**

SCHEME 2.3 Synthesis plans to make 1,3-diphenyl-2-(1-phenylethyl)-propane-1,3-dione via oxidative addition using phenylethane.

Gonzalez, A.; Marquet, J.; Moreno-Manas, M. *Tetrahedron Lett.* **1988**, 29, 1469

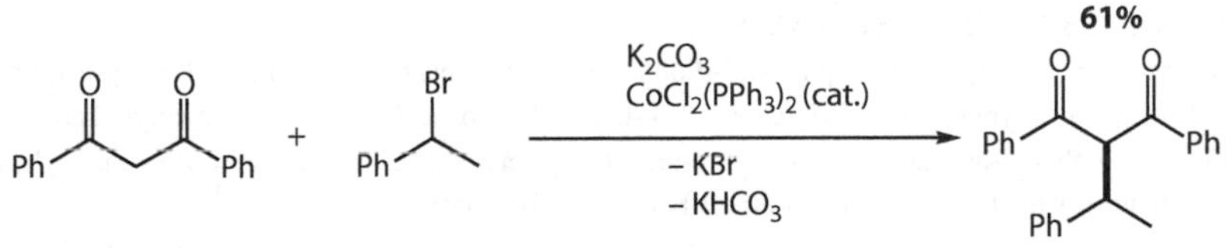

224 **184.9** **328**

Liu, J.; Wang, L.; Zheng, X.; Wang, A.; Zhu, M.; Yu, J.; Shen, Q. *Tetrahedron Lett.* **2012**, 53, 1843

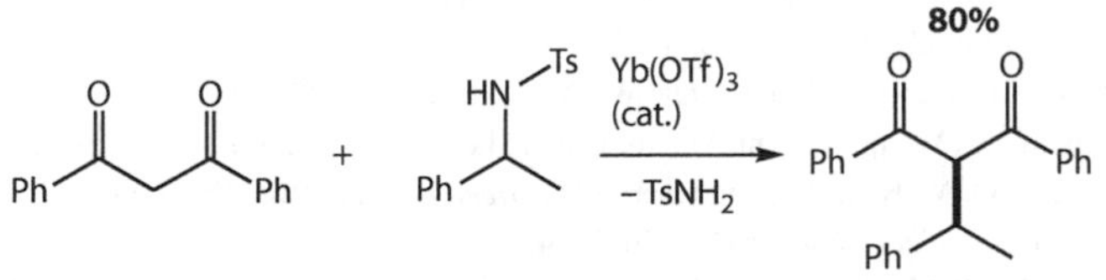

224 **275** **328**

SCHEME 2.4 Synthesis plans to make 1,3-diphenyl-2-(1-phenylethyl)-propane-1,3-dione using other reagents.

TABLE 2.1
Summary of Metrics for 1,3-Diphenyl-2-(1-Phenyl-Ethyl)-Propane-1,3-Dione Synthesis Plans[a]

Plan	Method[b]	% AE	% Yield	E-Kernel	E-Excess	E-Aux	E-Total	PMI
Das (2009)[c]	B	100	95	0.05	0.03	>0.3	>0.4	>1.4
Liu (2012)[d]	A conventional heating	94.8	98	0.07	1.4	0.05	1.5	2.5
Gonzalez (1988)[e]	D	60.0	61	1.7	2.1	>0.3	>4.1	>5.1
Liu (2012)[d]	A microwave heating	94.8	95	0.1	1.4	7.3	8.8	9.8
Liu (2007)[f]	A	94.8	92	0.1	0.4	>8.3	>8.8	>9.8
Li (2005)[g]	B	100	62	0.6	0.3	12.4	13.3	14.3
Wu (2008)[h]	A	94.8	99	0.06	0	>13.9	>14.0	>15.0
Liu (2007)[f]	B	100	52	0.9	0.7	>14.8	>16.4	>17.4
Wang, L. (2012)[i]	E	65.7	80	0.9	0.2	17.2	18.3	19.3
Liu (2010)[j]	A	94.8	99	0.06	1.4	>19.5	>20.9	>21.9
Li (2004)[k]	B	100	93	0.08	0.2	>22.1	>22.3	>23.3
Qureshi (2010)[l]	A	94.8	81	0.3	2.5	>73.5	>76.4	>77.4
Wang, G.W. (2008)[m]	B	100	94	0.06	0.8	>86.8	>87.7	>88.7
Merck-Frosst (2008)[n]	C	62.6	66	1.4	3.7	>318.9	>324.1	>325.1
Ramesh (2010)[o]	C	58.9	79	1.1	1.8	>858.9	>861.9	>862.9
Li (2007)[p]	B	100	63	0.6	0	1325.6	1326.2	1327.2

[a] Basis scale is 1 mole of target product.

[b] Method A = 1,3-Diphenyl-propane-1,3-dione + 1-phenyl-ethanol; method B = 1,3-diphenyl-propane-1,3-dione + styrene; method C = 1,3-diphenyl-propane-1,3-dione + phenylethane; method D = 1,3-diphenyl-propane-1,3-dione + 1-bromo-1-phenylethane; method E = 1,3-diphenyl-propane-1,3-dione + 4-methyl-N-(1-phenyl-ethyl)-benzenesulfonamide.

[c] Das, B., Krishnaiah, M., Laxminarayana, K., Damodar, K., and Kumar, D.N., *Chem. Lett.*, 2009, 38, 42.

[d] Xia, F., Le Zhao, Z., and Liu, P.N., *Tetrahedron Lett.*, 2012, 53, 2828.

[e] Gonzalez, A., Marquet, J., and Moreno-Manas, M., *Tetrahedron Lett.*, 1988, 29, 1469.

[f] Liu, P.N., Zhou, Z.Y., and Lau, C.P., *Chem. Eur. J.*, 2007, 13, 8610.

[g] Yao, X., and Li, C.J., *J. Org. Chem.*, 2005, 70, 5752.

[h] Li, Z., Duan, Z., Wang, H., Tian, R., Zhu, Q., and Wu, Y., *Synlett*, 2008, 2535.

[i] Liu, J., Wang, L., Zheng, X., Wang, A., Zhu, M., Yu, J., and Shen, Q., *Tetrahedron Lett.*, 2012, 53, 1843.

[j] Liu, P.N., Xia, F., Wang, Q.W., Ren, Y.J., and Chen, J.Q., *Green Chem.*, 2010, 12, 1049.

[k] Yao, X., and Li, C.J., *J. Am. Chem. Soc.*, 2004, 126, 6884.

[l] Qureshi, Z.S., Deshmukh, K.M., Tambade, P.J., and Bhanage, B.M., *Tetrahedron Lett.*, 2010, 51, 724.

[m] Wang, G.W., Shen, Y.B., Wu, X.L., and Wang, L., *Tetrahedron Lett.*, 2008, 49, 5090.

[n] Borduas, N., and Powell, D.A., *J. Org. Chem.*, 2008, 73, 7822.

[o] Ramesh, D., Ramulu, U., Rajaram, S., Prabhakar, P., and Venkateswarlu, Y., *Tetrahedron Lett.*, 2010, 51, 4898.

[p] Li, C.J., and Yao, X., *Org. Synth.*, 2007, 84, 222.

as only 5 out of 16 plans disclosed all materials appropriately. The plan described in *Organic Syntheses*, which utilized styrene, ranked the lowest but was the most honest in full disclosure of materials. Clearly, further improvements in that experimental procedure must focus on significantly reducing the solvent demand in chromatography or else choose recrystallization as the preferred method of purification since the product is a solid. Another point to address is the use of alternative catalysts that do not contain precious metals, such as gold or silver, which are obviously scarce and expensive. The plans using Amberlyst-15 (Das, 2009) or heteropolyacids such as $H_3PW_{12}O_{40}$ (Wang GW, 2008) would be recommended as greener and economically attractive catalysts for this transformation without sacrificing reaction performance.

EXAMPLE 2.2: 1-(4-PHENETHYL-PHENYL)-ETHANONE

This compound has been made by the following strategies: (1) Friedel–Crafts acylation (Scheme 2.5), (2) Suzuki cross-coupling (Scheme 2.6), (3) Molander modification of Suzuki cross-coupling (Scheme 2.7), (4) Hiyama cross-coupling (Scheme 2.8),

Meisters, A.; Wailes, P.C. *Aust. J. Chem.* **1966**, 19, 1215

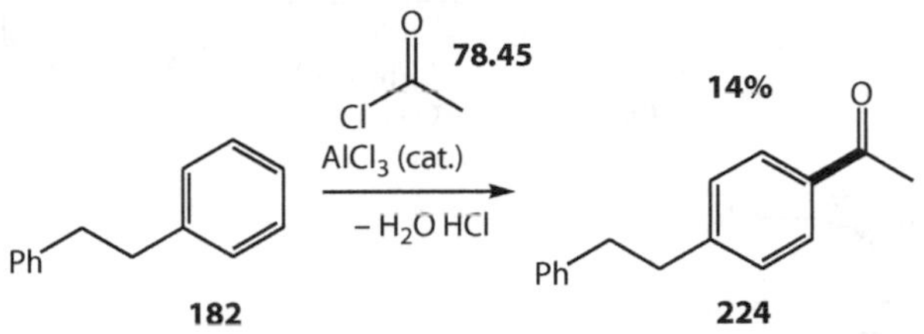

Lutz, R.E.; Allison, R.K.; Ashburn, G.; Bailey, P.S.; Clark, M.T.; Codington, J.F.; Deinet, A.J.; Freek, J.A.; Jordan, R.H.; Leake, N.H.; Martin, T.A.; Nicodemus, K.C.; Rowlett, R.J.; Shearer, N.H. Jr.; Smith, J.D.; Wilson, J.W.III *J. Org. Chem.* **1947**, 12, 617

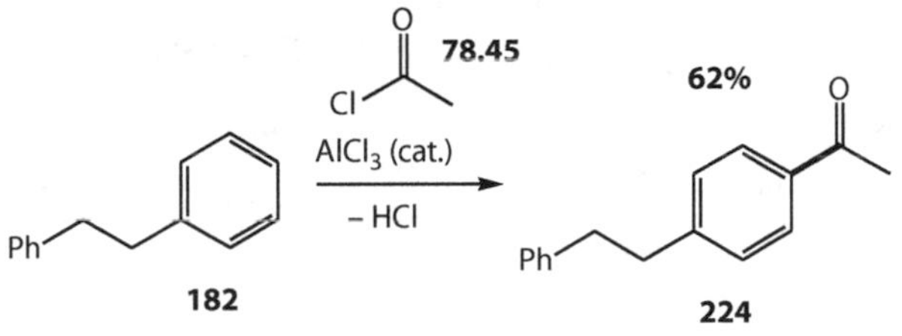

Boberg, F.; Muller, E.; Reddig, W. *J. Prakt. Chem.* **1995**, 337, 136

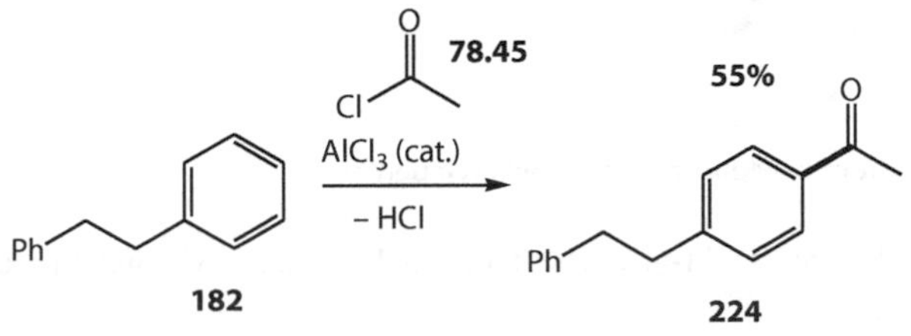

SCHEME 2.5   Synthesis of 1-(4-phenethyl-phenyl)-ethanone via Friedel–Crafts acylation.

Stepnicka, P.; Cisarova, I.; Schulz, J. *Organometallics* **2011**, 30, 4393

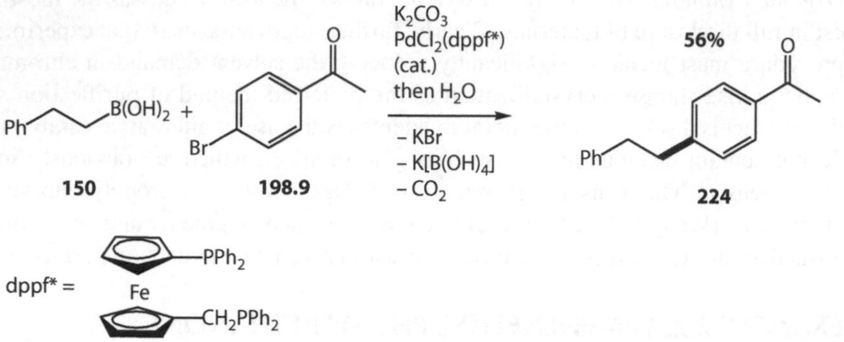

dppf* =

Molander, G.A.; Yun, C.S. *Tetrahedron* **2005**, 58, 1465

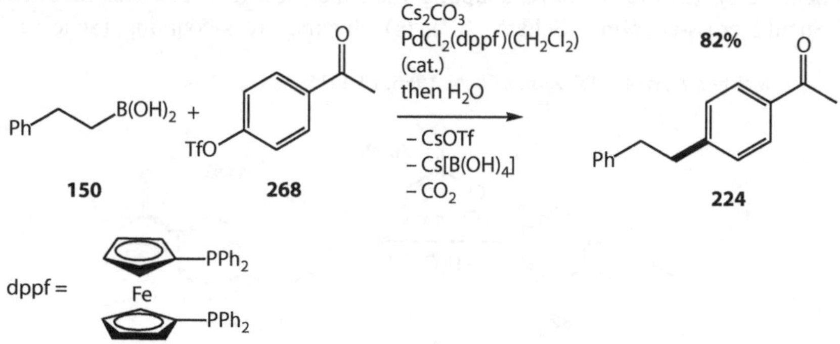

dppf =

Kondolff, I.; Doucet, H.; Santelli, M. *Tetrahedron* **2004**, 60, 3813

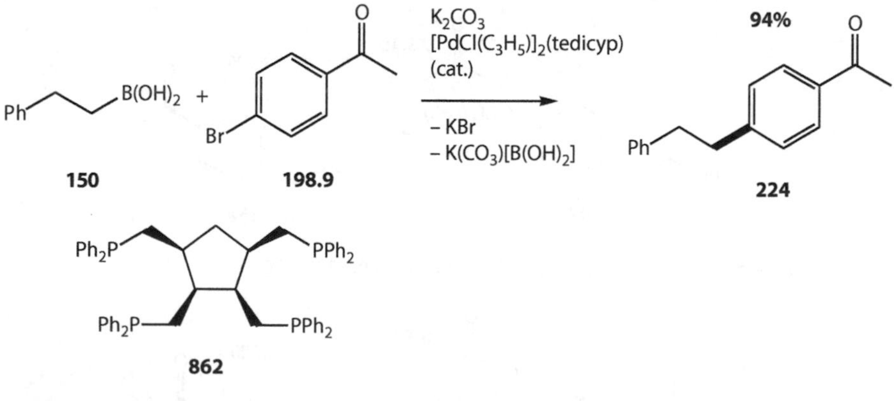

tedicyp = tetrakis(diphenylphosphinomethyl)cyclopentane

SCHEME 2.6   Synthesis of 1-(4-phenethyl-phenyl)-ethanone via Suzuki cross-coupling.

Molander, G.A.; Ito, T. *Org. Lett.* **2001**, 3, 393

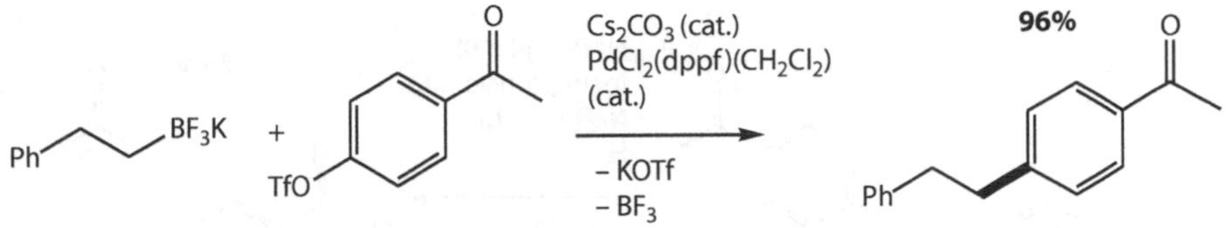

212   **268**   **96%**   **224**

Molander, G.A.; Petrillo, D.E. *Org. Synth.* **2007**, 84, 317

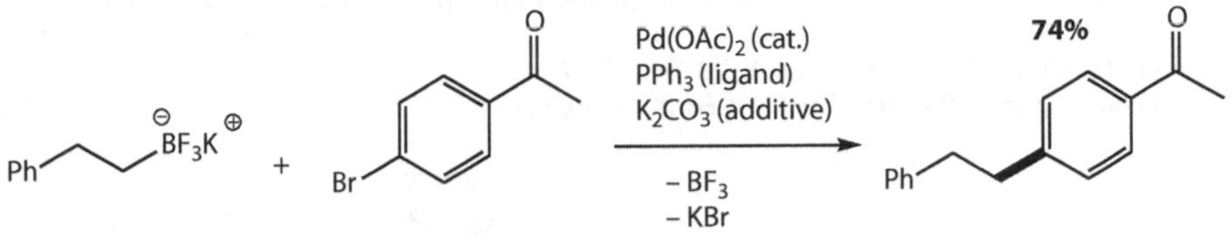

212   **198.9**   **74%**   **224**

SCHEME 2.7 Synthesis of 1-(4-phenethyl-phenyl)-ethanone via Molander modification to Suzuki cross-coupling.

Matsuhashi, H.; Asai, S.; Hirabayashi, K.; Hatanaka, Y.; Mori, A.; Hiyama, T. *Bull. Chem. Soc. Jpn.* **1997**, 70, 437

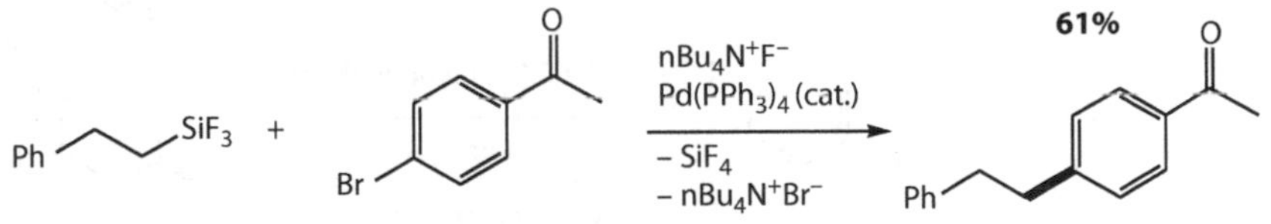

190   **198.9**   **61%**   **224**

Hatanaka, Y.; Hiyama, T. *Tetrahedron Lett.* **1990**, 31, 2719

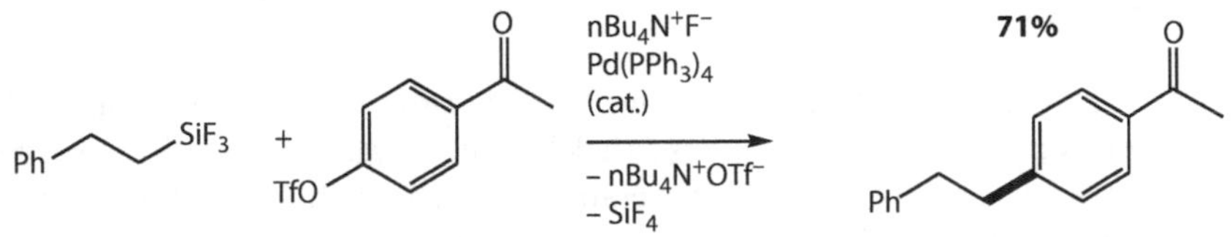

190   **268**   **71%**   **224**

Matsuhashi, H.; Kuroboshi, M.; Hatanaka, Y.; Hiyama, T. *Tetrahedron Lett.* **1994**, 35, 6507

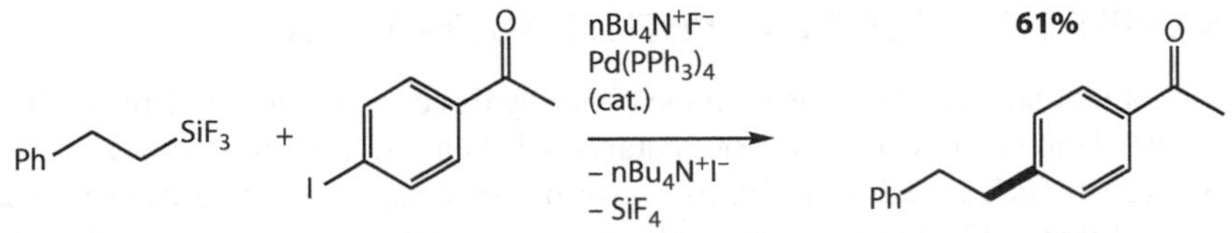

190   **246**   **61%**   **224**

SCHEME 2.8 Synthesis of 1-(4-phenethyl-phenyl)-ethanone via Hiyama cross-coupling.

Kantam, M.L.; Chakravarti, R.; Chintareddy, V.R.; Sreedhar, B.; Bhargava, S. *Adv. Synth. Catal.* **2008**, 350, 2544

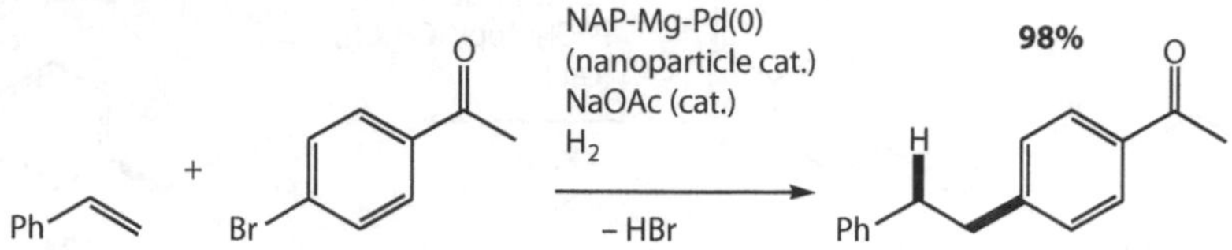

NAP-Mg-Pd(0) (nanoparticle cat.)
NaOAc (cat.)
H₂

98%

− HBr

104   198.9   224

SCHEME 2.9   Synthesis of 1-(4-phenethyl-phenyl)-ethanone via Heck cross-coupling.

Shen, Z.L.; Goh, K.K.K.; Yang, Y.S.; Lai, Y.C.; Wong, C.H.A.; Cheong, H.L.; Loh, T.P. *Angew. Chem. Int. Ed.* **2011**, 50, 511

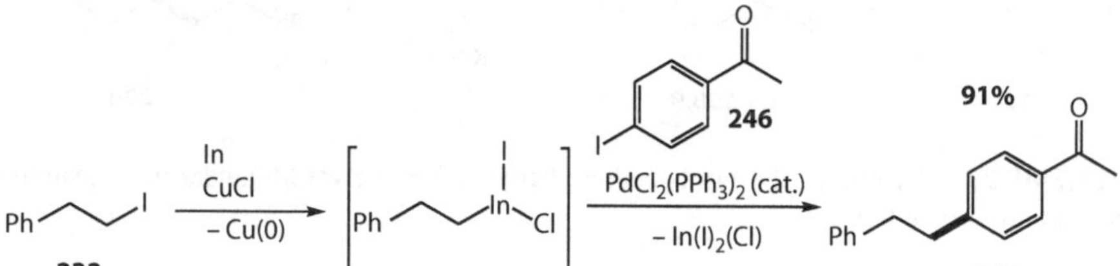

In
CuCl
− Cu(0)

246

PdCl₂(PPh₃)₂ (cat.)
− In(I)₂(Cl)

91%

232   224

Yan, C.S.; Peng, Y.; Xu, X.B.; Wang, Y.W. *Chem. Eur. J.* **2012**, 18, 6039

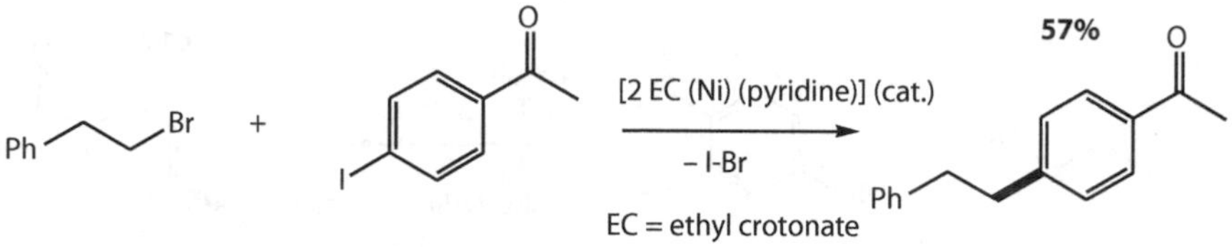

[2 EC (Ni) (pyridine)] (cat.)
− I-Br
EC = ethyl crotonate

57%

184.9   246   224

SCHEME 2.10   Synthesis of 1-(4-phenethyl-phenyl)-ethanone via Grignard-type cross-coupling.

(5) Heck cross-coupling (Scheme 2.9), and (6) Grignard-type reactions (Scheme 2.10). The most atom economical strategy is Friedel–Crafts acylation. Table 2.2 reveals that the Lutz 1947 plan has the lowest PMI overall, followed by the Boberg 1995 and Meisters 1966 plans, all of which follow the Friedel–Crafts protocol. Molander's 2007 *Organic Synthesis* procedure is the most reliable material-efficient cross-coupling strategy.

EXAMPLE 2.3: N-BENZYL-4-PHENYL-BUTYRAMIDE

This compound has largely been made either by direct amidation of 4-phenylbutyric acid with benzylamine using various catalysts (Scheme 2.11) or indirectly with sacrificial reagents such as carbonyldiimidazole or Burgess reagent to activate the carboxylic acid (Scheme 2.12). One odd example involved reaction of 1-chloro-4-phenyl-but-1-enyl acetate with two equivalents of benzylamine (Scheme 2.13). Table 2.3 indicates that direct amidation is the most atom economical strategy. Clearly, the use of

TABLE 2.2
Summary of Metrics for 1-(4-Phenethyl-Phenyl)-Ethanone Synthesis Plans[a]

Plan	Method[b]	% AE	% Yield	E-Kernel	E-Excess	E-Aux	E-Total	PMI
Lutz (1947)[c]	A	86.0	62	0.9	0.0004	8.0	8.9	9.9
Kondolff (2004)[d]	B	46.0	94	1.3	1.4	>8.2	>10.9	>11.9
Hiyama (1990)[e]	C	31.1	71	3.5	0.6	>18.7	>22.8	>23.8
Boberg (1995)[f]	A	86.0	55	1.1	0.05	31.2	32.3	33.3
Meisters (1966)[g]	A	86.0	14	7.2	0.04	28.6	35.8	36.8
Molander (2007)[h]	D	54.5	74	1.5	0.06	39.1	40.7	41.7
Hiyama (1994)[i]	C	32.1	61	4.1	5.2	>32.8	>42.1	>43.1
Hiyama (1997)[j]	C	34.5	61	3.7	7.1	>32.8	>43.7	>44.7
Shen (2011)[k]	E	32.4	91	2.4	2.4	>58.2	>63.0	>64.0
Stepnicka (2011)[l]	B	44.4	56	3.0	81.0	107.8	191.8	192.8
Kantam (2008)[m]	F	73.5	98	0.4	4.6	>307.7	>312.7	>313.7
Yan (2012)[n]	G	52.0	57	2.4	0.3	>775.5	>778.1	>779.1
Molander (2005)[o]	D	29.4	82	3.1	9.0	>1151.3	>1163.5	>1164.5
Molander (2001)[p]	D	46.7	96	1.2	0	>1485.7	>1486.9	>1487.9

[a] Basis scale is 1 mole of target product.

[b] Method A = Friedel–Crafts acylation; method B = Suzuki cross-coupling; method C = Hiyama cross-coupling; method D = Molander modification to Suzuki cross-coupling; method E = Grignard-type coupling using indium; method F = Heck cross-coupling; method G = Grignard-type coupling using nickel.

[c] Lutz, R.E. et al., *J. Org. Chem.,* 1947, 12, 617.

[d] Kondolff, I., Doucet, H., and Santelli, M., *Tetrahedron,* 2004, 60, 3813.

[e] Hatanaka, Y., and Hiyama, T., *Tetrahedron Lett.,* 1990, 31, 2719.

[f] Boberg, F., Muller, E., and Reddig, W., *J. Prakt. Chem.,* 1995, 337, 136.

[g] Meisters, A., and Wailes, P.C., *Aust. J. Chem.,* 1966, 19, 1215.

[h] Molander, G.A., and Petrillo, D.E., *Org. Synth.,* 2007, 84, 317.

[i] Matsuhashi, H., Kuroboshi, M., Hatanaka, Y., and Hiyama, T., *Tetrahedron Lett.,* 1994, 35, 6507.

[j] Matsuhashi, H., Asai, S., Hirabayashi, K., Hatanaka, Y., Mori, A., and Hiyama, T., *Bull. Chem. Soc. Jpn.,* 1997, 70, 437.

[k] Shen, Z.L., Goh, K.K.K., Yang, Y.S., Lai, Y.C., Wong, C.H.A., Cheong, H.L., and Loh, T.P., *Angew. Chem. Int. Ed.,* 2011, 50, 511.

[l] Stepnicka, P., Cisarova, I., and Schulz, J., *Organomet.,* 2011, 30, 4393.

[m] Kantam, M.L., Chakravarti, R., Chintareddy, V.R., Sreedhar, B., and Bhargava, S., *Adv. Synth. Catal.,* 2008, 350, 2544.

[n] Yan, C.S., Peng, Y., Xu, X.B., and Wang, Y.W., *Chem. Eur. J.,* 2012, 18, 6039.

[o] Molander, G.A., and Yun, C.S., *Tetrahedron,* 2005, 58, 1465.

[p] Molander, G.A., and Ito, T., *Org. Lett.,* 2001, 3, 393.

Verma, R.; Ghosh, S.K. *J. Perkin Trans. I* **1998**, 2377

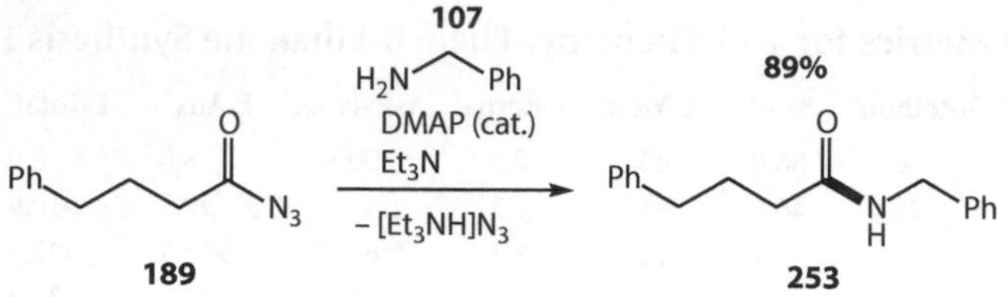

Maki, T.; Ishihara, K.; Yamamoto, H. *Tetrahedron* **2007**, 63, 8645

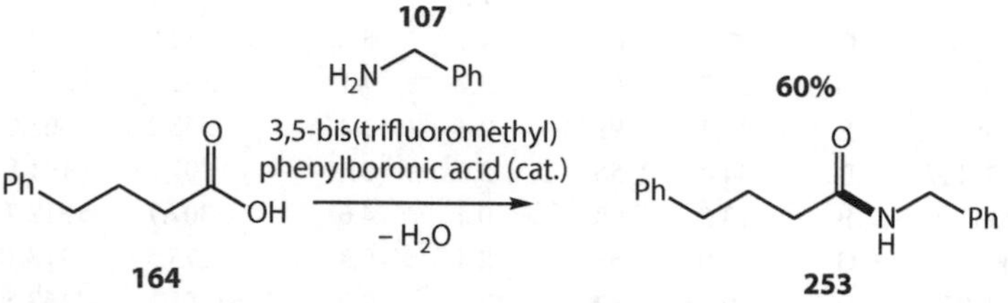

Maki, T.; Ishihara, K.; Yamamoto, H. *Org. Lett.* **2006**, 8, 1431

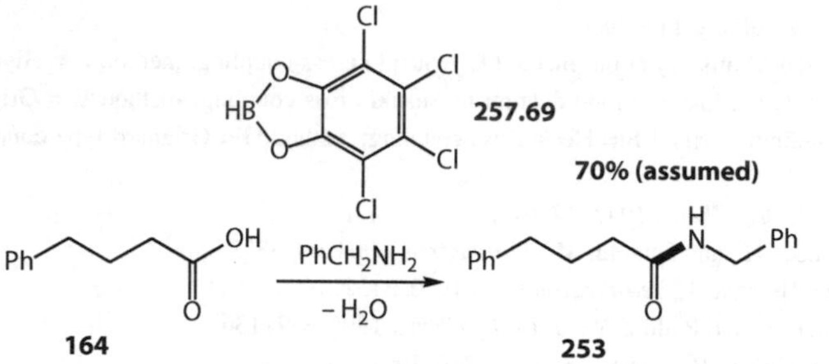

Hall, D.; Gernigon, N.; Zoubi, R.; Thornton, P.D. *WO* 2012109749 (**2012**)

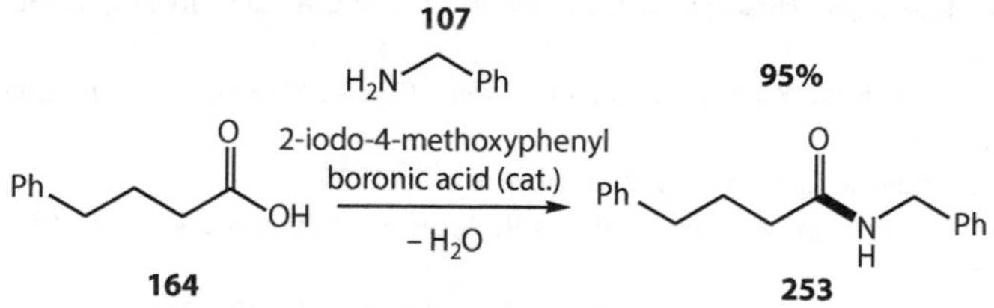

Tang, P. *Org. Synth.* **2005**, 81, 262

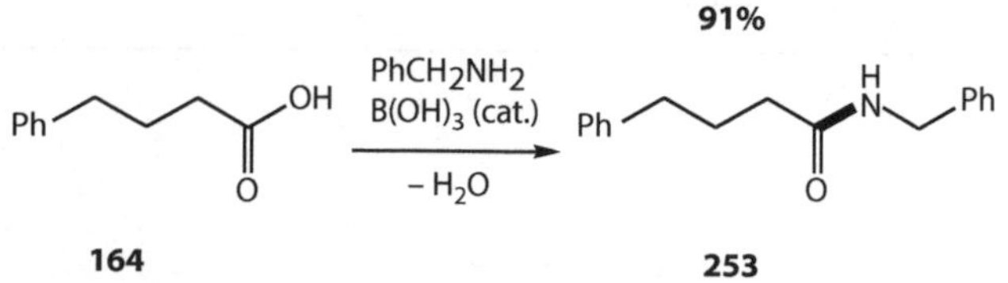

SCHEME 2.11   Synthesis of N-benzyl-4-phenyl-butyramide via catalyzed amidation.

Metro, T.X.; Bonnamour, J.; Reidon, T.; Sarpoulet, J.; Martinez, J.; Lamaty, F. *Chem. Commun.* **2012**, 11781

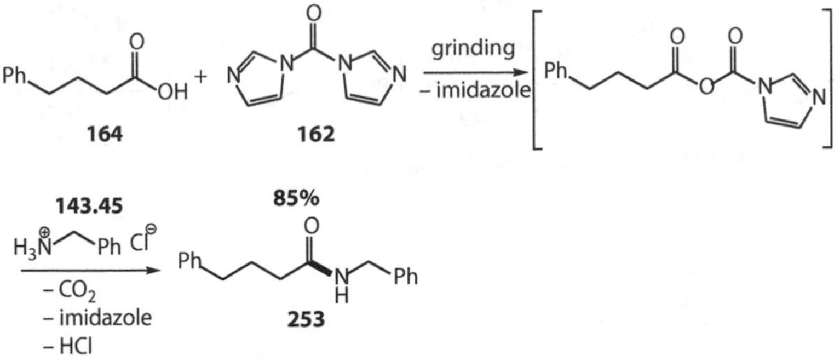

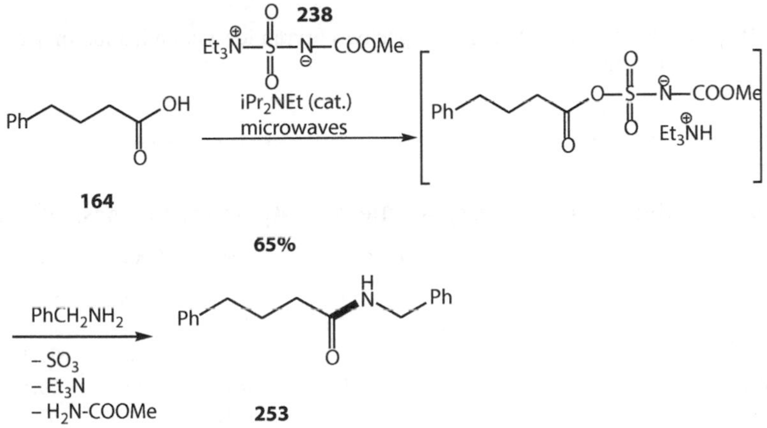

SCHEME 2.12   Synthesis of N-benzyl-4-phenyl-butyramide via amidation using sacrificial reagents. (From Bejot, R. et al., *Eur. J. Org. Chem.*, 101, 2007.)

sacrificial reagents will significantly lower the performance of this metric. However, the procedure with the least PMI does indeed use the sacrificial activating reagent carbonyldiimidazole. The reason that this plan rates highest is because it was carried out without reaction solvent and implemented mechanochemical activation. The second best PMI protocol used the other sacrificial Burgess reagent under microwave conditions with a minimum amount of acetonitrile solvent. The *Organic Syntheses* procedure (Tang, 2007) used a significant volume of workup extraction solvents, which weighed down the overall PMI. Clearly, reducing these would make that direct boric acid-catalyzed amidation more competitive with the former two. This example is a nice tutorial illustration of how a highly atom economical strategy may be jeopardized by the use of excessive volumes of solvent. True optimization is achieved when the best plan both is highly atom economical *and* uses the least volume of solvent. Such a strategy constitutes orchestrated optimization where all material efficiency metrics are synergistically improved in the same direction.

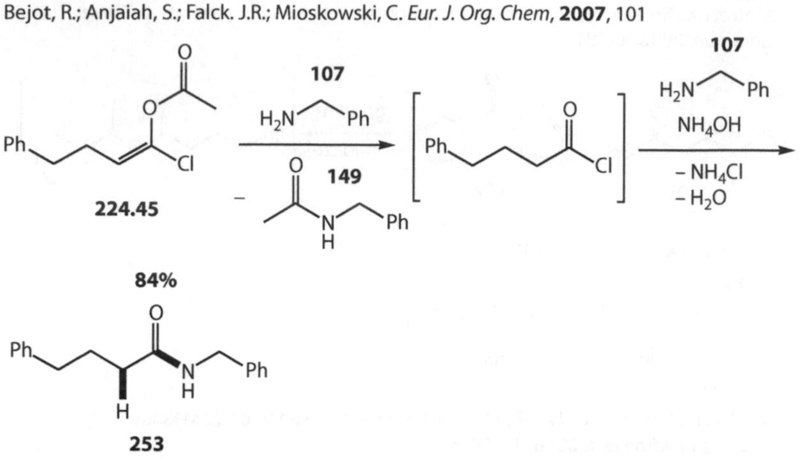

Bejot, R.; Anjaiah, S.; Falck. J.R.; Mioskowski, C. *Eur. J. Org. Chem*, **2007**, 101

SCHEME 2.13 Synthesis of N-benzyl-4-phenyl-butyramide via amidation of 1-chloro-4-phenyl-but-1-enyl acetate.

TABLE 2.3
Summary of Metrics for N-Benzyl-4-Phenyl-Butyramide Synthesis Plans[a]

Plan	No. Steps	% AE	% Yield	E-Kernel	E-Excess	E-Aux	E-Total	PMI
Metro (2012)[b]	1	53.9	85	1.2	0.2	48.1	49.5	50.5
Makara (2006)[c]	1	59.7	65	2.1	1.8	57.9	61.8	62.8
Falck (2007)[d]	1	53.4	84	1.2	9.8	>66.3	>77.3	>78.3
Hall (2012)[e]	1	93.4	95	0.1	0.07	86.4	86.6	87.6
Tang (2005)[f]	1	93.4	91	0.2	0.02	87.7	87.8	88.8
Maki (2006)[g]	1	93.4	80	0.5	0	>115.7	>116.2	>117.2
Maki (2007)[h]	1	93.4	60	0.8	0	>144.1	>144.9	>145.9
Ghosh (1998)[i]	1	63.7	89	0.8	1.4	>351.4	>353.6	>354.6

[a] Basis scale is 1 mole of target product.

[b] Metro, T.X., Bonnamour, J., Reidon, T., Sarpoulet, J., Martinez, J., and Lamaty, F., *Chem. Commun.*, 2012, 11781.

[c] Wodka, D., Robbins, M., Lan, P., Martinez, R.L., Athanasopoulos, J., and Makara, G.M., *Tetrahedron Lett.*, 2006, 47, 1825.

[d] Bejot, R., Anjaiah, S., Falck, J.R., and Mioskowski, C., *Eur. J. Org. Chem.*, 2007, 101.

[e] Hall, D., Gernigon, N., Zoubi, R., and Thornton, P.D., WO 2012109749 (2012).

[f] Tang, P., *Org. Synth.*, 2005, 81, 262.

[g] Maki, T., Ishihara, K., and Yamamoto, H., *Org. Lett.*, 2006, 8, 1431.

[h] Maki, T., Ishihara, K., and Yamamoto, H., *Tetrahedron*, 2007, 63, 8645.

[i] Verma, R., and Ghosh, S.K., *J. Perkin Trans. I*, 1998, 2377.

EXAMPLE 2.4: CYCLEN

Cyclen is a 12-membered macrocycle that is made by various multistep strategies beginning from N,N′-bis-(2-amino-ethyl)-ethane-1,2-diamine (triethylene tetraamine) to make either sacrificial 5-6-5 tricyclics that are reduced to the product (Scheme 2.14) or sacrificial 5-6-6-5 quadricyclics that are then hydrolyzed or reduced to the desired product (Scheme 2.15). Other strategies employed are multicomponent cycloaddition reaction motifs beginning from small molecules to make 5-6-6-5 quadricyclic intermediates that are then hydrolyzed or oxidized to

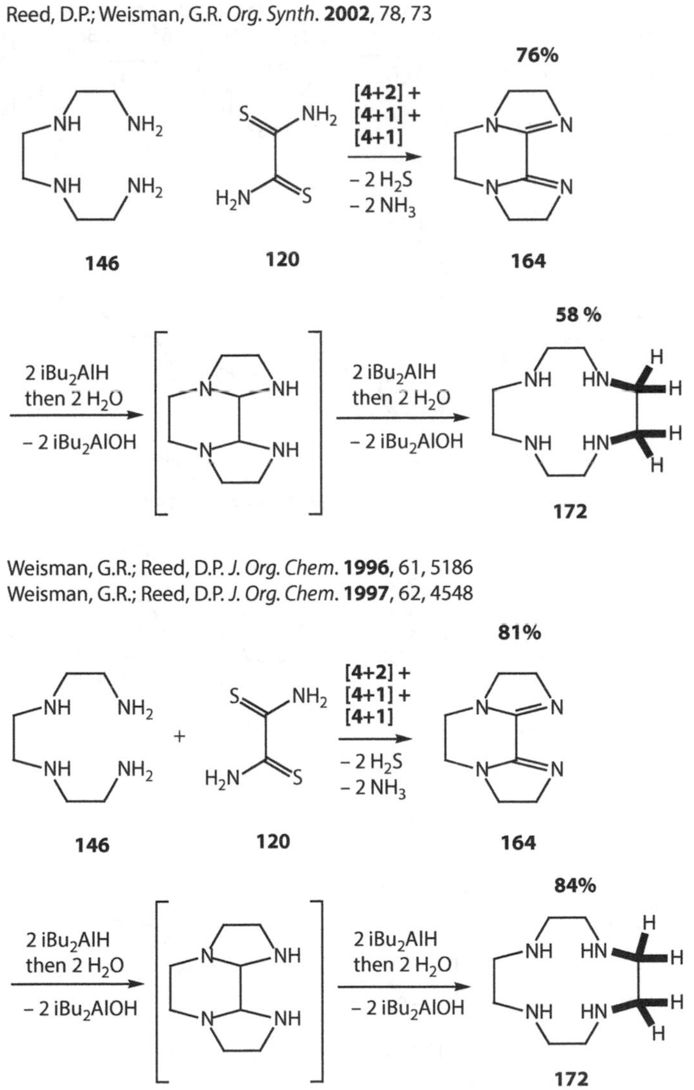

Reed, D.P.; Weisman, G.R. Org. Synth. 2002, 78, 73
76%
[4+2] + [4+1] + [4+1]
- 2 H2S
- 2 NH3
146
120
164
58 %
2 iBu2AlH then 2 H2O
- 2 iBu2AlOH
172
Weisman, G.R.; Reed, D.P. J. Org. Chem. 1996, 61, 5186
Weisman, G.R.; Reed, D.P. J. Org. Chem. 1997, 62, 4548
81%
146
120
164
84%
172

SCHEME 2.14 Synthesis of cyclen via 5-6-5 tricyclic intermediates.

the desired product (Scheme 2.16). Table 2.4 shows that the top two performing plans are the Dow Chemical and Sandnes patents. The former, with the lowest overall atom economy, involves a [2+2+2] + [2+2+2] cycloaddition to make a 5-6-6-5 quadricyclic intermediate, which is then hydrolyzed under basic conditions (see entry 1 in Scheme 2.16). The latter, with the highest overall atom economy,

Sandnes, R.W.; Gacek, M.; Undheim, K. *Acta Chem. Scand.* **1998**, 52, 1402

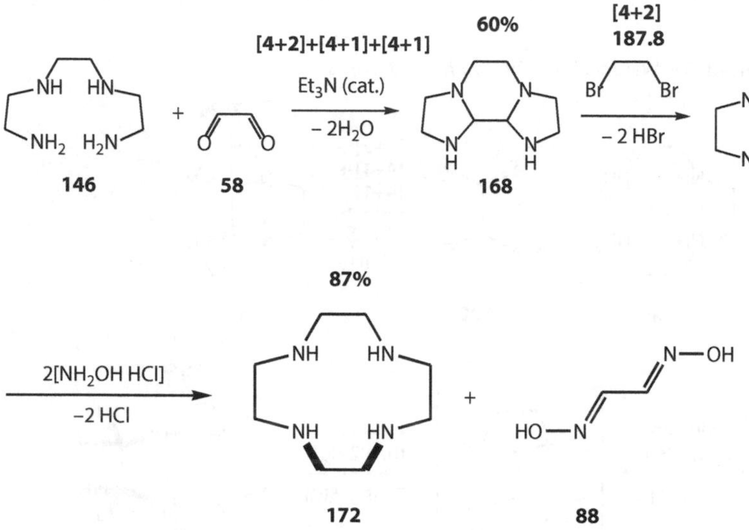

Sandnes, R.W.; Vasilevskis, J.; Undheim, K.; Gacek, M. WO 9628432 (**1996**)

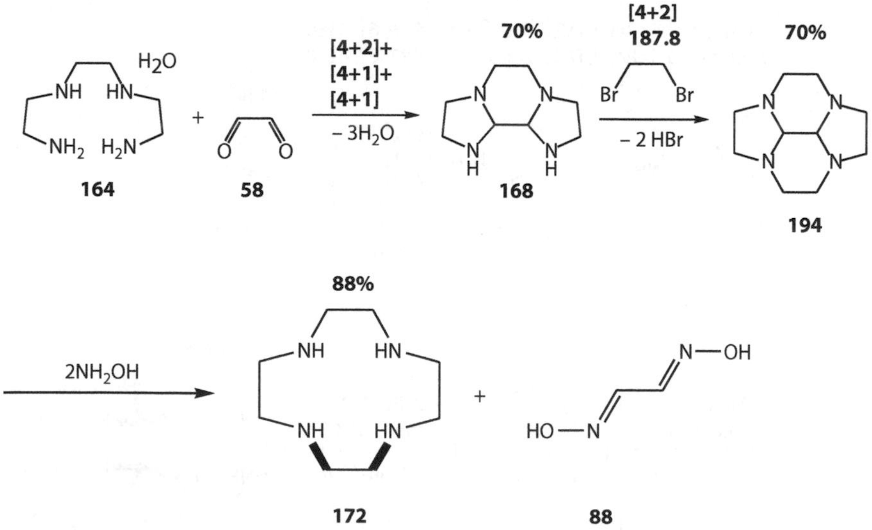

SCHEME 2.15 Synthesis of cyclen via 5-6-6-5 quadricyclic intermediates. (Top from Herve, G. et al., *Eur. J. Org. Chem.*, 33, 2000.)

Herve, G.; Bernard, H.; Toupet, L.; Handel, H. *Eur. J. Org. Chem.* **2000**, 33

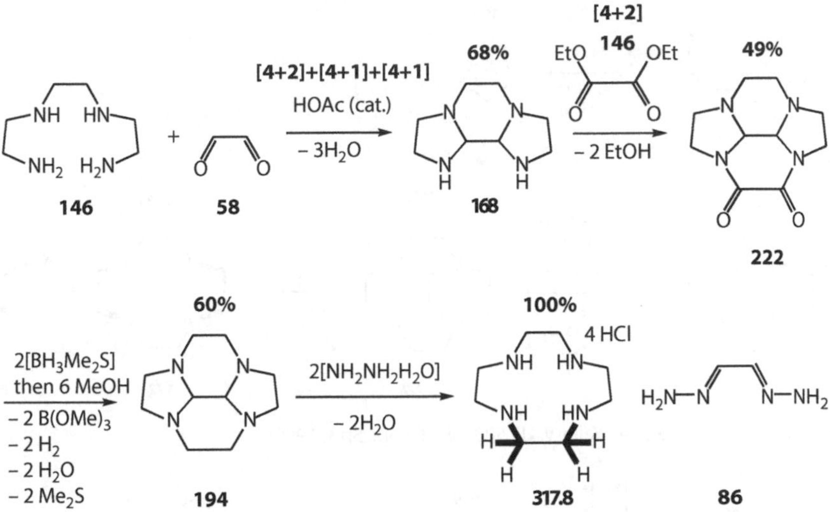

Athey, P.S.; Kiefer, G.E. *J. Org. Chem.* **2002**, 67, 4081

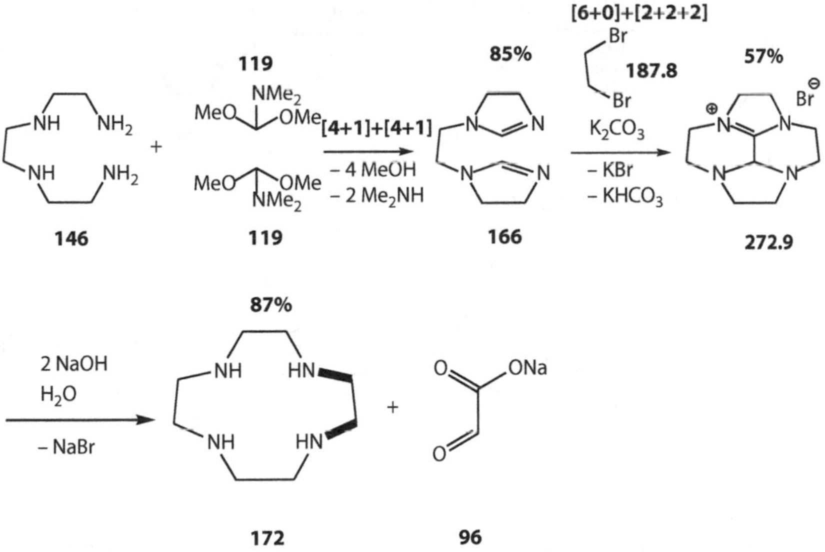

SCHEME 2.15 (Continued)

involves a sequential [4+2] + [4+1] + [4+1] cycloaddition to a 5-6-5 tricyclic intermediate, which then reacts with 1,2-dibromoethane in a [4+2] fashion to make a 5-6-6-5 quadricyclic intermediate whose central two-carbon framework is excised with hydroxylamine hydrochloride (see entry 2 in Scheme 2.15). The *Organic Syntheses* plan (Reed, 2002) ranks lowest in PMI yet has the highest reported overall yield. Again, the culprit is the excessive use of auxiliary materials in workup and purification.

Athey, P.S.; Kiefer, G.E. US 5587451 (Dow Chemical, 1996)

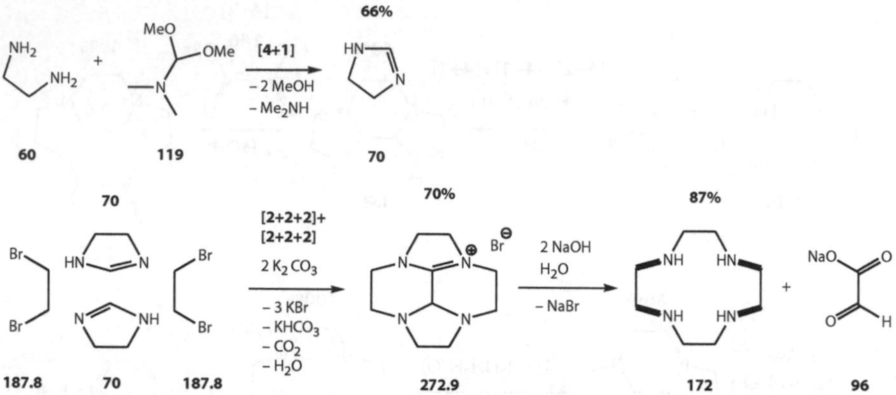

Argese, M.; Ripa, G.; Scala, A.; Valle, V. US 5880281 (Dibra, SpA, 1999)

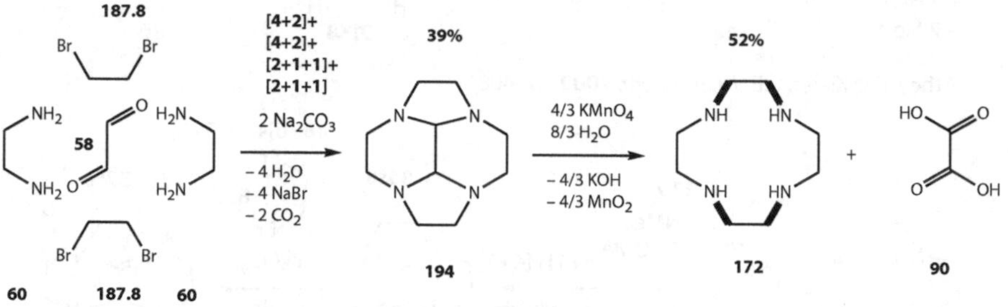

Ripa, G.; Argese, M. US 5886174 (Dibra SpA, 1999)

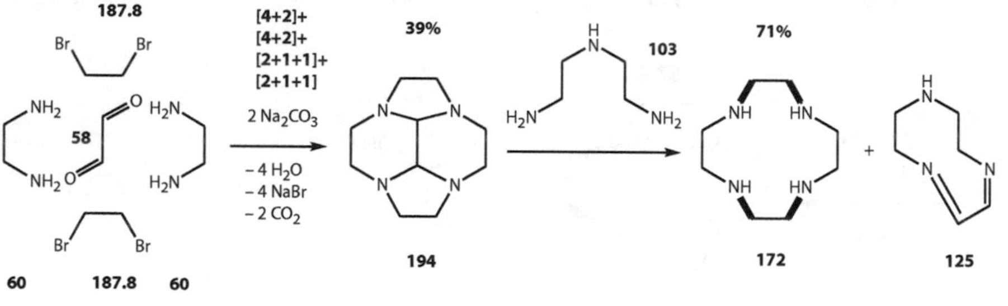

SCHEME 2.16 Synthesis of cyclen via multicomponent cycloadditions to 5-6-6-5 quadricyclic intermediates.

TABLE 2.4
Summary of Metrics for Cyclen Synthesis Plans[a]

Plan	No. Steps	% AE	% Yield	E-Kernel	E-Excess	E-Aux	E-Total	PMI
Dow (1996)[c]	3	15.5	40.2	11.1	12.3	20.5	43.9	44.9
Sandnes (1996)[d]	3	36.1	43.1	4.2	1.7	47.2	53.2	54.2
Dow (2002)[e]	3	21.3	42.2	8.8	11.4	107.2	127.4	128.4
Sandnes (1998)[f]	3	32.4	21.4	8.5	4.7	119.4	132.7	133.7
Dibra G2 (1999)[g]	2	19.8	27.7	15.9	19.9	133.6	169.2	170.2
Dibra G1 (1999)[h]	2	16.8	20.3	23.8	37.5	174.9	236.3	237.3
Reed (1996)[i]	2	19.0	68.0	5.7	6.4	246.1	258.3	259.3
Herve (2000)[b,j]	4	33.8	20.0	6.4	1.4	522.5	530.2	531.2
Reed (2002)[k]	2	19.0	44.1	8.9	67.9	580.2	657.0	658.0

[a] Basis scale is 1 mole of target product.
[b] Target compound is tetrahydrochloride salt.
[c] Athey, P.S., and Kiefer, G.E., U.S. 5587451, Dow Chemical, 1996.
[d] Sandnes, R.W., Vasilevskis, J., Undheim, K.,and Gacek, M., WO 9628432, 1996.
[e] Athey, P.S., and Kiefer, G.E., *J. Org. Chem.,* 2002, 67, 4081.
[f] Sandnes, R.W., Gacek, M., and Undheim, K., *Acta Chem. Scand.,* 1998, 52, 1402.
[g] Ripa, G., and Argese, M., U.S. 5886174, Dibra SpA, 1999.
[h] Argese, M., Ripa, G., Scala, A., and Valle, V., U.S. 5880281, Dibra, SpA, 1999.
[i] Weisman, G.R., and Reed, D.P., *J. Org. Chem.,* 1996, 61, 5186.
[j] Herve, G., Bernard, H., Toupet, L., and Handel, H., *Eur. J. Org. Chem.,* 2000, 33.
[k] Reed, D.P., and Weisman, G.R., *Org. Synth.,* 2002, 78, 73.

EXAMPLE 2.5: DIBENZYL-(1-CYCLOHEXYL-3-TRIMETHYLSILANYL-PROP-2-YNYL)-AMINE

Three-component coupling of an aldehyde, alkyne, and amine has been used to make racemic (Scheme 2.17) and optically active versions (Scheme 2.18) of this compound. The stereospecific (R) and (S) isomers were made using (S)-quinap and (R)-quinap as chiral auxiliaries, respectively. Tables 2.5 and 2.6 show the plan rankings for each product type. In both cases the *Organic Syntheses* procedures are most reliable since all materials were disclosed, particularly in chromatographic procedures. Furthermore, since the same group reported all plans using the same set of reaction conditions, there is little diversity of protocols to make further judgments about plan efficiencies.

EXAMPLE 2.6: [2-(4-FLUORO-PHENYL)-1H-INDOL-4-YL]-PYRROLIDIN-1-YL-METHANONE

The Merck synthesis of this pharmaceutical candidate by a three-step sequence is shown in Scheme 2.19. Table 2.7 shows that the *Organic Syntheses* procedure produces less than half the waste compared to the original report. Again, the large savings came as a consequence of reducing the solvent demand in auxiliary material usage.

Dube, H.; Gommermann, N.; Knochel, P. *Synthesis* **2004**, 2015

197

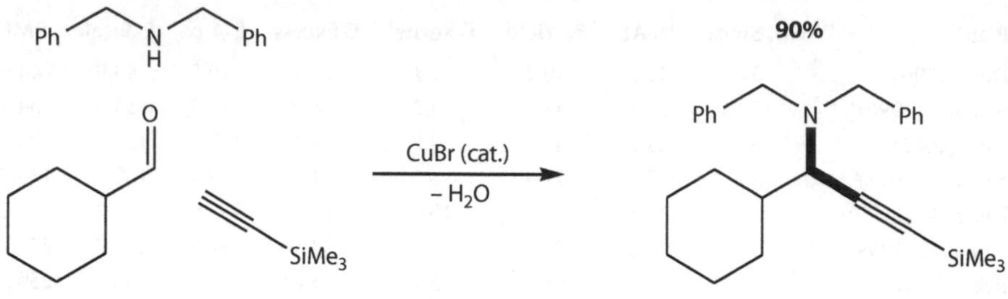
Ph
N
H
Ph
O
SiMe3
CuBr (cat.)
– H2O
90%
Ph
N
Ph
SiMe3

112 **98** **389**

Gommermann, N.; Knochel, P. *Tetrahedron* **2005**, 61, 11418

197

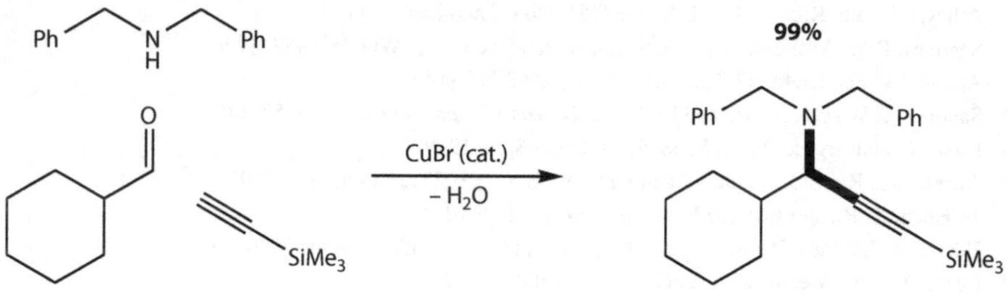
Ph
N
H
Ph
O
SiMe3
CuBr (cat.)
– H2O
99%
Ph
N
Ph
SiMe3

112 **98** **389**

Gommermann, N.; Knochel, P. *Org. Synth.* **2007**, 84, 1

197

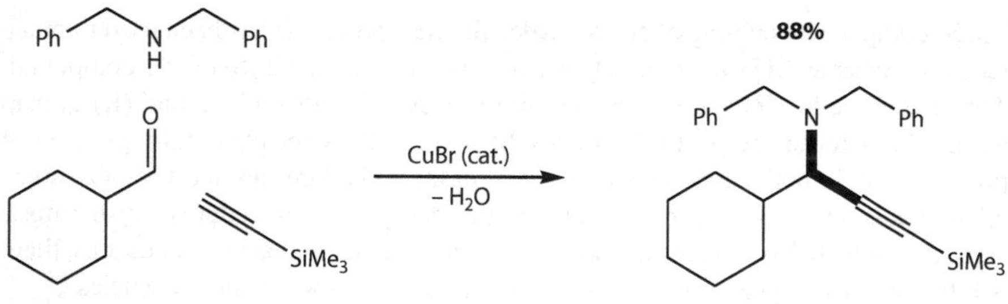
Ph
N
H
Ph
O
SiMe3
CuBr (cat.)
– H2O
88%
Ph
N
Ph
SiMe3

112 **98** **389**

SCHEME 2.17 Synthesis of (±)-dibenzyl-(1-cyclohexyl-3-trimethylsilanyl-prop-2-ynyl)-amine.

Gommermann, N.; Knochel, P. *Tetrahedron* **2005**, 61, 11418

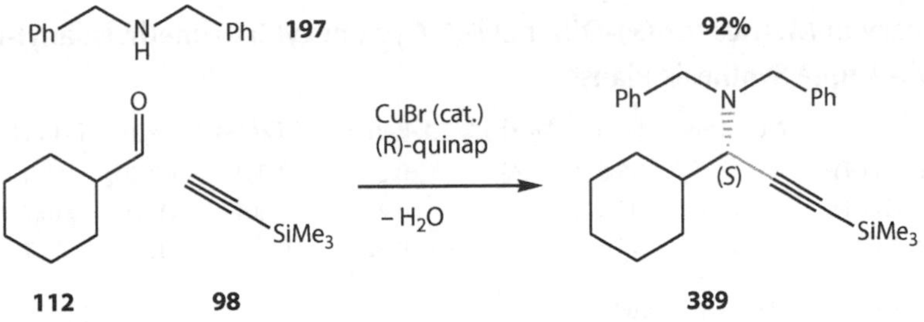

Gommermann, N.; Knochel, P. *Chem. Commun.* **2004**, 2324

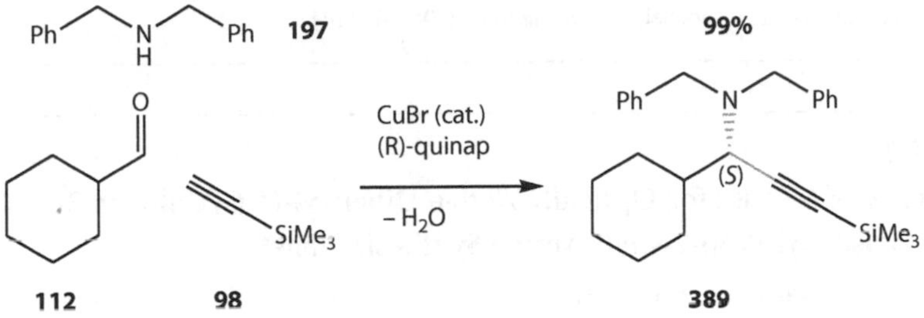

Gommermann, N.; Knochel, P. *Org. Synth.* **2007**, 84, 1

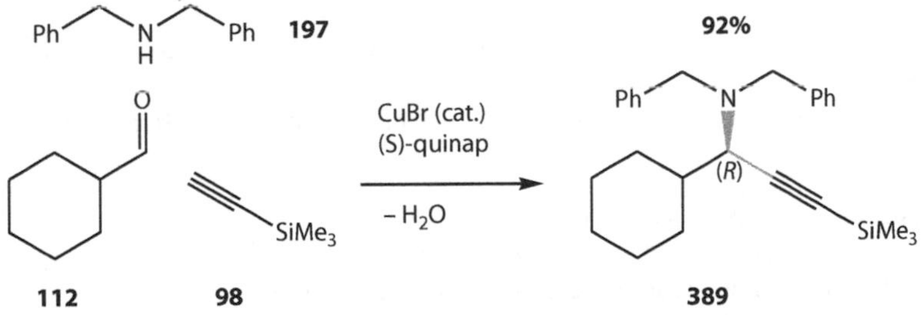

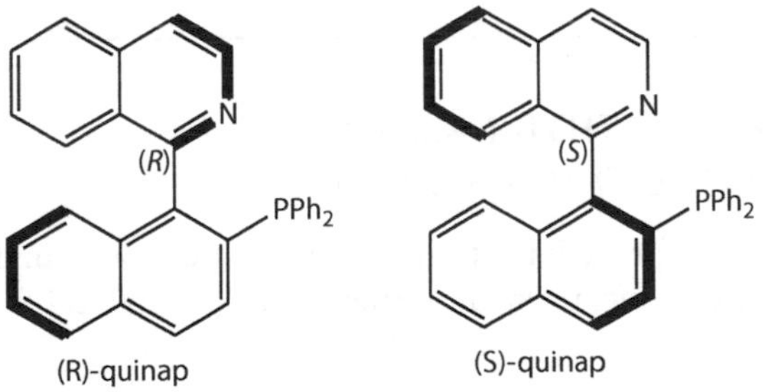

SCHEME 2.18 Synthesis of optically active dibenzyl-(1-cyclohexyl-3-trimethylsilanyl-prop-2-ynyl)-amine.

TABLE 2.5

Summary of Metrics for (±)-Dibenzyl-(1-Cyclohexyl-3-Trimethylsilanyl-Prop-2-ynyl)-Amine Synthesis Plans[a]

Plan	No. Steps	% AE	% Yield	E-Kernel	E-Excess	E-Aux	E-Total	PMI
Knochel (2007)[b]	1	95.6	88	0.2	0.001	188.2	188.4	189.4
Knochel (2004)[c]	1	95.6	90	0.2	0.001	>10.5	>10.7	>11.7
Knochel (2005)[d]	1	95.6	99	0.06	0	>13.7	>13.7	>14.7

[a] Basis scale is 1 mole of target product.
[b] Gommermann, N., and Knochel, P., *Org. Synth.,* 2007, 84, 1.
[c] Dube, H., Gommermann, N., and Knochel, P., *Synthesis,* 2004, 2015.
[d] Gommermann, N., and Knochel, P., *Tetrahedron,* 2005, 61, 11418.

TABLE 2.6

Summary of Metrics for Optically Active Dibenzyl-(1-Cyclohexyl-3-Trimethylsilanyl-Prop-2-ynyl)-Amine Synthesis Plans[a]

Plan	Product Stereoisomer	No. Steps	% AE	% Yield	E-Kernel	E-Excess	E-Aux	E-Total	PMI
Knochel (2007)[b]	R	1	95.6	92	0.1	0.0009	390.3	390.4	391.4
Knochel (2004)[c]	S	1	95.6	99	0.06	0	>6.6	>6.7	>7.7
Knochel (2005)[d]	S	1	95.6	92	0.1	0	>80.1	>80.2	>81.2

[a] Basis scale is 1 mole of target product.
[b] Gommermann, N., and Knochel, P., *Org. Synth.,* 2007, 84, 1.
[c] Dube, H., Gommermann, N., and Knochel, P., *Synthesis,* 2004, 2015.
[d] Gommermann, N., and Knochel, P., *Tetrahedron,* 2005, 61, 11418.

EXAMPLE 2.7: 3-HYDROXYMETHYL-4-METHYL-2-METHYLENE-PENTANOIC ACID METHYL ESTER

Paquette's *Organic Syntheses* procedure to make the title compound (see Scheme 2.20) is a modest improvement over an earlier report (see Table 2.8). Both strategies were identical.

EXAMPLE 2.8: (3-PHENYL-ALLYL)-PHOSPHINIC ACID

The title compound was made by direct coupling of phenylallene and phosphinic acid as shown in Scheme 2.21. The *Organic Syntheses* procedure using $Pd(OAc)_2$ instead of $Pd_2(dba)_3$ as catalyst and employing less than half the volume of auxiliary materials boosted the overall material efficiency performance for this compound (see Table 2.9).

Kuethe, J.T.; Davies, I.W. *Tetrahedron* **2006**, 62, 11381

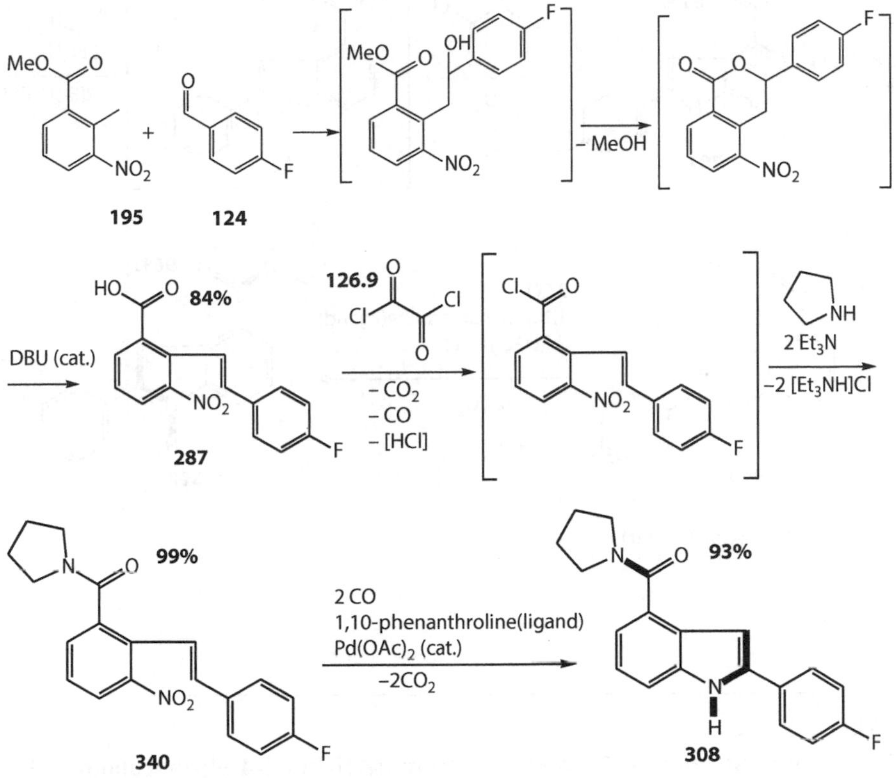

Kuethe, J.T.; Beutner, G.L. *Org. Synth.* **2009**, 86, 92

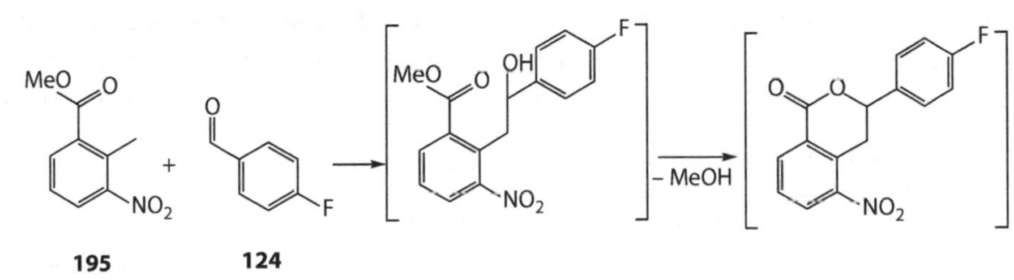

SCHEME 2.19 Synthesis of [2-(4-fluoro-phenyl)-1H-indol-4-yl]-pyrrolidin-1-yl-methanone.

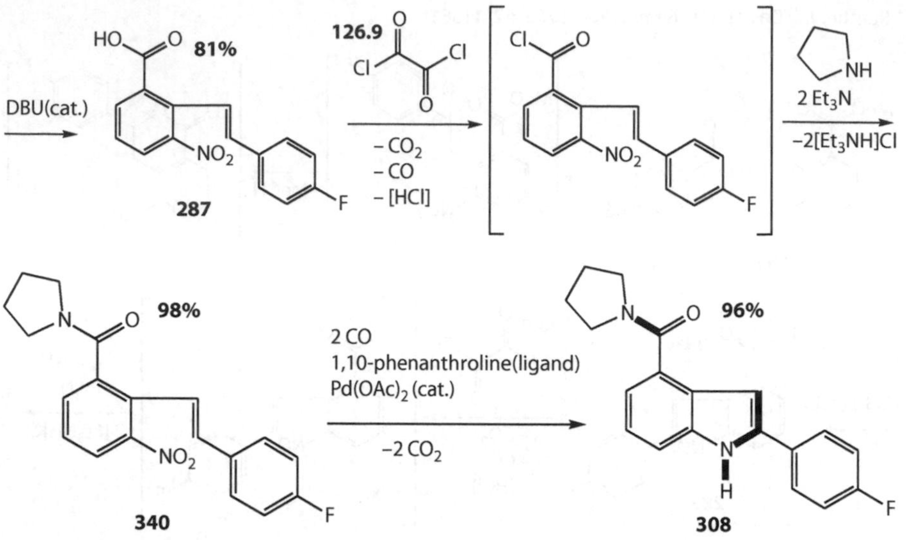

SCHEME 2.19 (*Continued*)

TABLE 2.7

Summary of Metrics for [2-(4-Fluoro-Phenyl)-1H-Indol-4-yl]-Pyrrolidin-1-yl-Methanone Synthesis Plans[a]

Plan	No. Steps	% AE	% Yield	E-Kernel	E-Excess	E-Aux	E-Total	PMI
Merck (2009)[b]	3	39.7	76.2	1.9	0.6	272.4	274.9	275.9
Merck (2006)[c]	3	39.7	77.3	1.9	0.4	606.1	608.5	609.5

[a] Basis scale is 1 mole of target product.
[b] Kuethe, J.T., and Beutner, G.L., *Org. Synth.*, 2009, 86, 92.
[c] Kuethe, J.T., and Davies, I.W., *Tetrahedron*, 2006, 62, 11381.

Paquette, L.A.; Bennett, G.D.; Isaac, M.B.; Chhatriwalla, A. *J. Org. Chem.* **1998**, 63, 1836

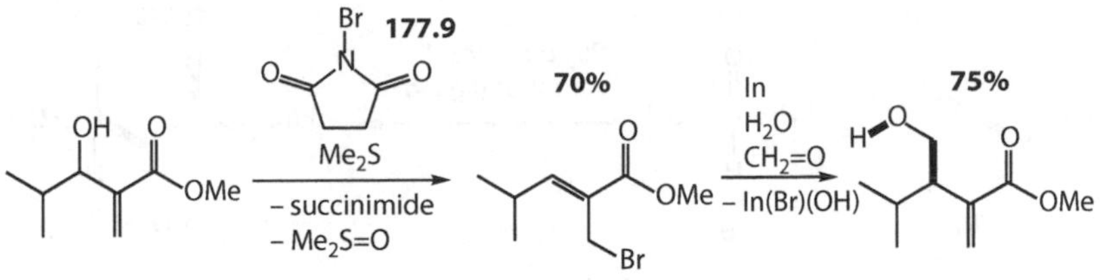

158 220.9 172

Bennett, G.D.; Paquette, L.A. *Org. Synth.* **2000**, 77, 107

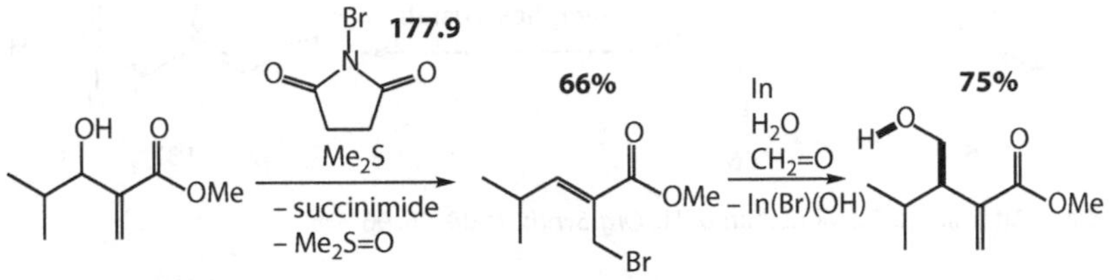

158 220.9 172

SCHEME 2.20 Synthesis of 3-hydroxymethyl-4-methyl-2-methylene-pentanoic acid methyl ester.

TABLE 2.8
Summary of Metrics for 3-Hydroxymethyl-4-Methyl-2-Methylene-Pentanoic Acid Methyl Ester Synthesis Plans[a]

Plan	No. Steps	% AE	% Yield	E-Kernel	E-Excess	E-Aux	E-Total	PMI
Paquette (2000)[b]	2	30.7	49.5	4.9	18.4	303.4	326.7	327.7
Paquette (1998)[c]	2	30.7	52.5	4.7	36.3	316.9	357.9	358.9

[a] Basis scale is 1 mole of target product.
[b] Bennett, G.D., and Paquette, L.A., *Org. Synth.*, 2000, 77, 107.
[c] Paquette, L.A., Bennett, G.D., Isaac, M.B., and Chhatriwalla, A., *J. Org. Chem.*, 1998, 63, 1836.

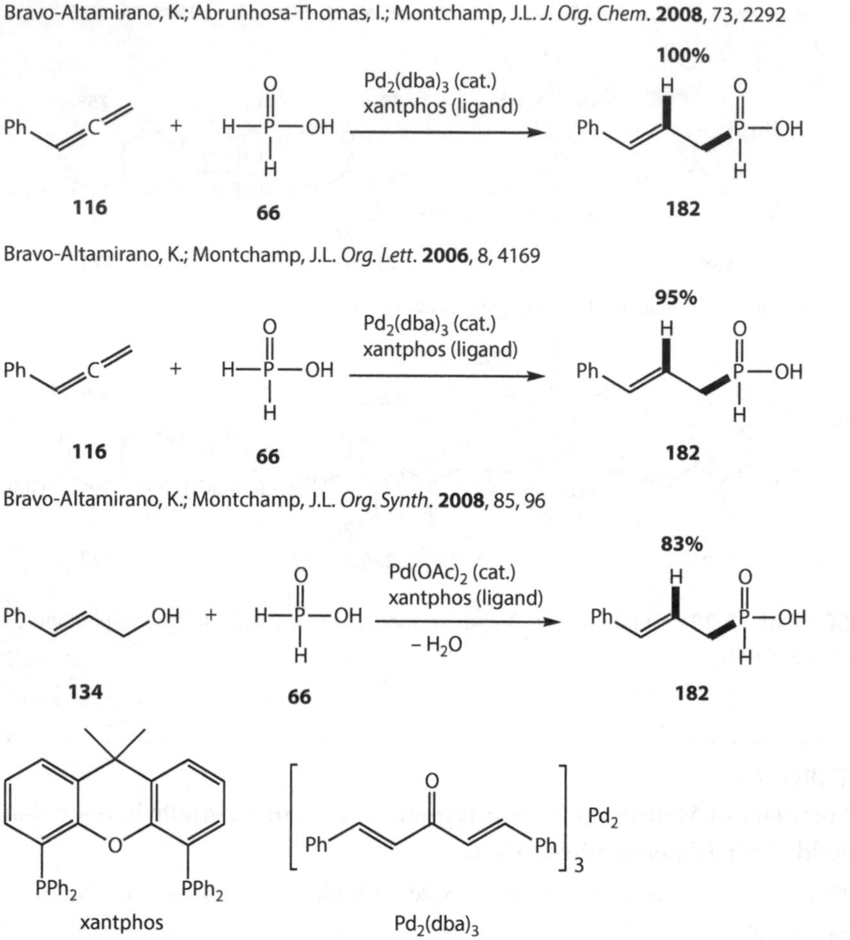

Bravo-Altamirano, K.; Abrunhosa-Thomas, I.; Montchamp, J.L. *J. Org. Chem.* **2008**, 73, 2292

Bravo-Altamirano, K.; Montchamp, J.L. *Org. Lett.* **2006**, 8, 4169

Bravo-Altamirano, K.; Montchamp, J.L. *Org. Synth.* **2008**, 85, 96

SCHEME 2.21 Synthesis plans for (3-phenyl-allyl)-phosphinic acid from phenylallene and phosphinic acid.

TABLE 2.9
Summary of Metrics for (3-Phenyl-Allyl)-Phosphinic Acid Synthesis Plans[a]

Plan	No. Steps	% AE	% Yield	E-Kernel	E-Excess	E-Aux	E-Total	PMI
Montchamp (2008)[b]	1	100	95	0.3	0.4	80.0	80.8	81.8
Montchamp (2008)[c]	1	100	100	0	0.4	131.4	131.8	132.8
Montchamp (2006)[d]	1	100	83	0.2	0.02	184.1	184.2	185.2

[a] Basis scale is 1 mole of target product.
[b] Bravo-Altamirano, K., and Montchamp, J.L., *Org. Synth.*, 2008, 85, 96.
[c] Bravo-Altamirano, K., Abrunhosa-Thomas, I., and Montchamp, J.L., *J. Org. Chem.*, 2008, 73, 2292.
[d] Bravo-Altamirano, K., and Montchamp, J.L., *Org. Lett.*, 2006, 8, 4169.

EXAMPLE 2.9: 9-PROPIONYL-3,4,5,8,9,10-HEXAHYDRO-OXECIN-2-ONE

A comparison of the synthesis of the title compound in a two-step process with and without isolation of the bicyclic lactol intermediate as shown in Scheme 2.22 is a nice demonstration of reduced waste production when two steps are coupled together (see Table 2.10).

EXAMPLE 2.10: 3A-HYDROXY-2,3-DIISOPROPOXY-4, 6A-DIMETHYL-4,5,6,6A-TETRAHYDRO-3AH-PENTALEN-1-ONE

The three-component coupling of diisopropyl squarate with two equivalents of 2-bromopropene is an efficient strategy to assemble the [3.3.0] bicyclic ring system of the title compound as shown in Scheme 2.23. The optimized *Organic Synthesis* procedure produces about 50 % less waste because fewer auxiliary materials are used (see Table 2.11).

Webb, K.S.; Asirvatham, E.; Posner, G.H. *Org. Synth. Coll. Vol.* 8, 562 (**1993**)

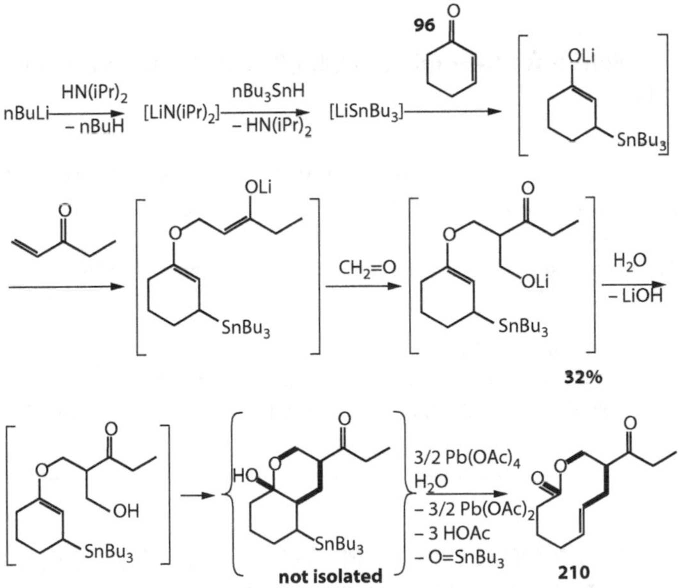

Posner, G.H.; Webb, K.S.; Asirvatham, E.; Jew, S.; Degl'Innocenti, A. *J. Am. Chem. Soc.* **1988**, 110, 4754

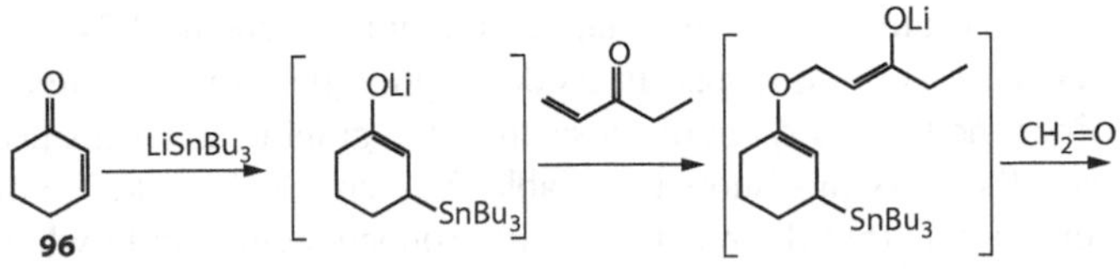

SCHEME 2.22 Synthesis plans for the macrolide 9-propionyl-3,4,5,8,9,10-hexahydro-oxecin-2-one.

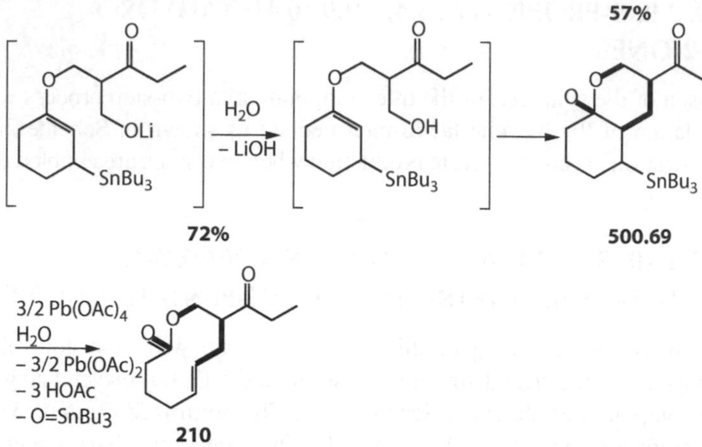

SCHEME 2.22 (Continued)

TABLE 2.10
Summary of Metrics for 9-Propionyl-3,4,5,8,9,10-Hexahydro-Oxecin-2-One Synthesis Plans[a]

Plan	No. Steps	% AE	% Yield	E-Kernel	E-Excess	E-Aux	E-Total	PMI
Posner (1993)[b]	2 (without isolation of intermediate)	16.6	32	17.9	42.0	668.9	728.8	729.8
Posner (1988)[c]	2 (with isolation of intermediate)	17.4	41	9.6	265.8	626.2	901.6	902.6

[a] Basis scale is 1 mol of target product.
[b] Webb, K.S., Asirvatham, E., and Posner, G.H., *Org. Synth. Coll.*, 8, 562, 1993.
[c] Posner, G.H., Webb, K.S., Asirvatham, E., Jew, S., and Degl'Innocenti, A., *J. Am. Chem. Soc.*, 110, 4754, 1988.

EXAMPLE 2.11: 7,7-DIMETHYL-3-PHENYL-4-P-TOLYL-1,4,6,7,8,9-HEXAHYDRO-PYRAZOLO[3,4-B]QUINOLIN-5-ONE

The [3+2+1] coupling of dimedone, 4-methylbenzaldehyde, and 5-phenyl-2H-pyrazol-3-ylamine via microwave irradiation is shown in Scheme 2.24. The *Organic Syntheses* procedure is more honest in its description than the two other literature procedures since the latter did not disclose chromatographic materials used in purification. Hence the rankings shown in Table 2.12 should be taken with caution. Another observation is that the reaction yield dropped significantly when the scale of the reaction was increased four-fold.

Morwick, T.; Paquette, L.A. *Org. Synth. Coll. Vol.* 9, 670 (1998)
Paquette, L.A.; Morwick, T.M. *J. Am. Chem. Soc.* **1997**, 119, 1230

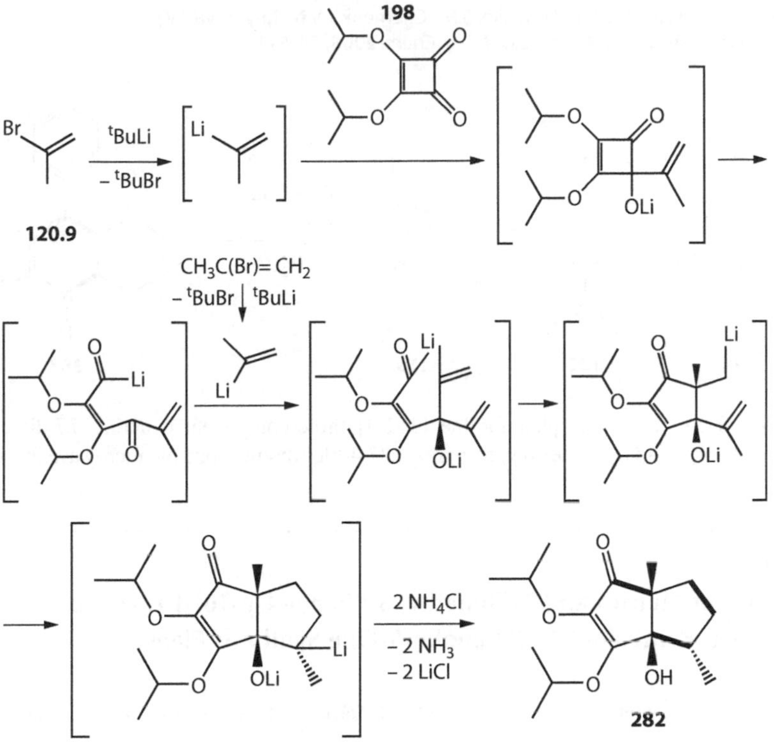

SCHEME 2.23 Synthesis plan for the polyquinane 3a-hydroxy-2,3-diisopropoxy-4,6a-dimethyl-4,5,6,6a-tetrahydro-3aH-pentalen-1-one via a [4+1]+[2+2+1] three component cyclization.

TABLE 2.11

Summary of Metrics for 3a-Hydroxy-2,3-Diisopropoxy-4,6a-Dimethyl-4,5,6,6a-Tetrahydro-3aH-Pentalen-1-One Synthesis Plans[a]

Plan	No. Steps	% AE	% Yield	E-Kernel	E-Excess	E-Aux	E-Total	PMI
Paquette (1998)[b]	1	41.8	75	2.2	7.9	317.4	327.5	328.5
Paquette (1997)[c]	1	41.8	90	1.7	7.9	642.8	652.4	653.4

[a] Basis scale is 1 mol of target product.
[b] Morwick, T., and Paquette, L.A., *Org. Synth. Coll.*, 9, 670, 1998.
[c] Paquette, L.A., and Morwick, T.M., *J. Am. Chem. Soc.*, 119, 1230, 1997.

Glasnov, T.N.; Kappe, C.O. *Org. Synth.* **2009**, 86, 252

Andriushchenko, A.Y.; Desenko, S.M.; Chernenko, V.N.; Chebanov, V.A. *J. Heterocyclic Chem.* **2011**, 48, 365

Chebanov, V.A.; Saraev, V.E.; Desenko, S.M.; Chernenko, V.N.; Knyazeva, I.V.; Groth, U.; Glasnov, T.N.; Kappe, C.O. *J. Org. Chem.* **2008**, 73, 5110

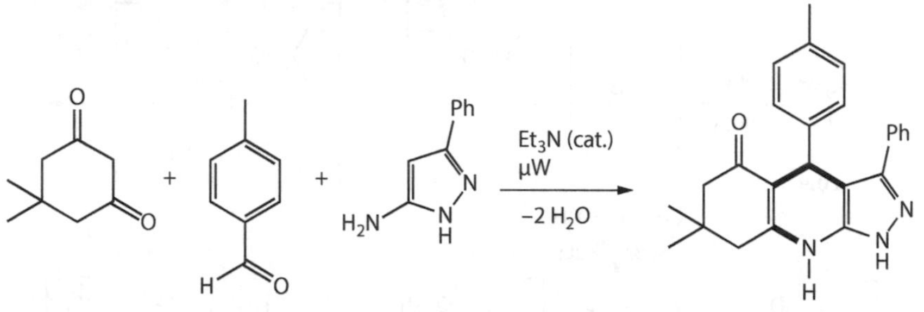

140   **120**   **159**   **383**

SCHEME 2.24   Synthesis plan for the [3+2+1] three-component coupling 7,7-dimethyl-3-phenyl-4-p-tolyl-1,4,6,7,8,9-hexahydro-pyrazolo[3,4-b]quinolin-5-one by microwave irradiation.

TABLE 2.12

Summary of Metrics for 7,7-Dimethyl-3-Phenyl-4-p-Tolyl-1,4,6,7,8,9-Hexahydro-Pyrazolo[3,4-b]Quinolin-5-One Synthesis Plans[a]

Plan	No. Steps	% AE	% Yield	E-Kernel	E-Excess	E-Aux	E-Total	PMI
Kappe (2008)[b]	1	91.4	86	0.3	0	>22.5	>22.8	>23.8
Chebanov (2011)[c]	1	91.4	83	0.3	0	>26.3	>26.6	>27.6
Kappe (2009)[d]	1	91.4	50	1.2	0.002	445.3	446.5	447.5

[a] Basis scale is 1 mol of target product.

[b] Chebanov, V.A., Saraev, V.E., Desenko, S.M., Chernenko, V.N., Knyazeva, I.V., Groth, U., Glasnov, T.N., and Kappe, C.O., *J. Org. Chem.*, 73, 5110, 2008.

[c] Andriushchenko, A.Y., Desenko, S.M., Chernenko, V.N., and Chebanov, V.A., *J. Heterocyclic Chem.*, 48, 365, 2011.

[d] Glasnov, T.N., and Kappe, C.O., *Org. Synth.*, 86, 252, 2009.

REFERENCES

1. *Organic Syntheses* website, http://www.orgsyn.org/Instructions_to_authors.pdf, http://www.orgsyn.org/AuthorChecklist.pdf (accessed June 2013).

2. (a) O'Neil, M.J., Smith, A., Heckelman, P.E., eds. *The Merck Index: An Encyclopedia of Chemicals, Drugs, and Biologicals*, 13th ed. Merck & Co., Rahway, NJ, 2001. (b) Mundy, B.P., Ellerd, M.G. *Name Reactions and Reagents in Organic Synthesis.* Wiley, New York, 1988. (c) Laue, T., Plagens, A. *Named Organic Reactions.* Wiley, New York, 2000. (d) Smith, J.G., Fieser, M., Fieser and Fieser's *Reagents for Organic Synthesis Collective Index for Volumes 1–12.* Wiley. New York, 1990. (e) Li, J.J.

Name Reactions: A Collection of Detailed Reaction Mechanisms. Springer-Verlag, Berlin, 2002. (f) Andraos, J. Named Things in Chemistry and Physics. http://www.careerchem.com/NAMED/Homepage.html (accessed June 2013).

3. Wöhler, F. *Ann. Chim.* 1828, 37, 330.
4. (a) Julian, P.L., Pikl, J. *J. Am. Chem. Soc.* 1935, 57, 563. (b) Julian, P.L., Pikl, J. *J. Am. Chem. Soc.* 1935, 57, 755. (c) Julian, P.L., Pikl, J. *J. Am. Chem. Soc.* 1935, 57, 539. (d) Julian, P.L., Pikl, J., Boggess, D. *J. Am. Chem. Soc.* 1934, 56, 1797.
5. Anastas, P.T., Warner, J.C. *Green Chemistry: Theory and Practice.* Oxford University Press, New York, 1998.
6. Jiménez-González, C., Ponder, C.S., Broxterman, Q.B., Manley, J.B. *Org. Process Res. Dev.* 2011, 15, 912.
7. Danheiser, R.L. *Org. Synth.* 2011, 88, 1.
8. (a) Andraos, J. *Org. Process Res. Dev.* 2009, 13, 161. (b) Andraos, J. *The Algebra of Organic Synthesis: Green Metrics, Design Strategy, Route Selection, and Optimization.* CRC Press, Boca Raton, 2012.

3 Reaction 1: Synthesis of 3-Benzyl-5-Methyleneoxazolidin-2-One from N-Benzylprop-2-yn-1-Amine and CO_2

Qing-Wen Song and Liang-Nian He

CONTENTS

INTRODUCTION

Carbon dioxide (CO_2) has drawn much attention from the viewpoint of global warming as well as one of the most promising alternatives to phosgene.[1] The chemical conversion of CO_2 is a viable approach to utilize an abundant and inexpensive C1 feedstock.[1,2] The preparation of oxazolidinones is one of the most attractive synthetic methods from CO_2. Much effort has been made to develop the chemical fixation of CO_2 into propargylic amines to provide the oxazolidinone derivatives (Scheme 3.1) because of oxazolidinones showing a wide application as intermediates[3] and chiral auxiliaries[4] in organic synthesis. A green effective catalytic process

47

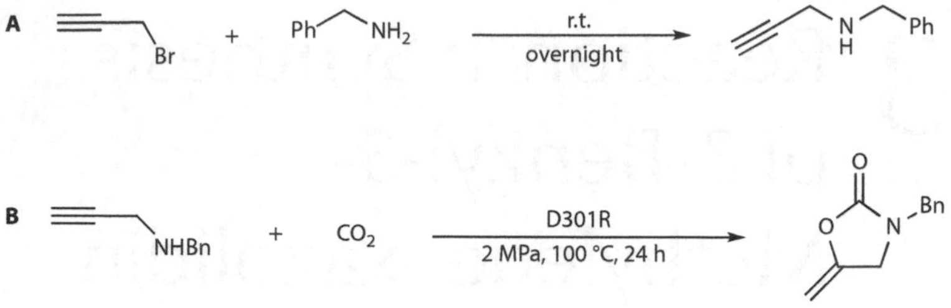

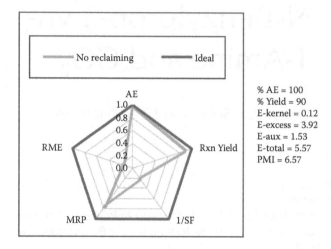

SCHEME 3.1 General protocols for the synthesis of N-benzylprop-2-yn-1-amine and 3-benzyl-5-methylencoxazolidin-2-one.

% AE = 100
% Yield = 90
E-kernel = 0.12
E-excess = 3.92
E-aux = 1.53
E-total = 5.57
PMI = 6.57

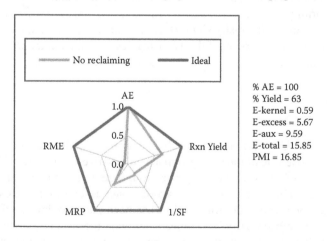

FIGURE 3.1 Radial pentagon for synthesis of 3-benzyl-5-methyleneoxazolidin-2-one (this work). (Note: Metrics do not include chromatographic materials and preparation of catalysts.)

% AE = 100
% Yield = 63
E-kernel = 0.59
E-excess = 5.67
E-aux = 9.59
E-total = 15.85
PMI = 16.85

FIGURE 3.2 Radial pentagon for synthesis of 3-benzyl-5-methyleneoxazolidin-2-one. (Note: Metrics do not include chromatographic materials and preparation of catalysts.) (See Mitsudo, T. et al., *Tetrahedron Lett.*, 28, 4417–4418, 1987 [7].)

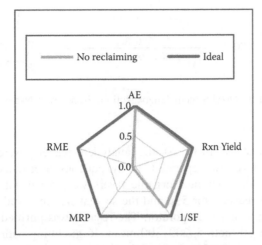

% AE = 100
% Yield = 95
E-kernel = 0.052
E-excess = 0.25
E-aux = 18.41
E-total =18.71
PMI = 19.71

FIGURE 3.3 Radial pentagon for synthesis of 3-benzyl-5-methyleneoxazolidin-2-one. (Note: Metrics do not include chromatographic materials and preparation of catalysts.) (See Kikuchi, S. et al., *Bull. Chem. Soc. Jpn.*, 84, 698–717, 2011 [8].)

TABLE 3.1
Summary of Green Metrics for Synthesis of
3-Benzyl-5-Methyleneoxazolidin-2-One[a]

Plan	% AE[b]	% Yield	E-Kernel	E-Excess	E-Aux	E-Total	PMI[c]
He	**100**	**90**	**0.12**	**3.92**	**1.53**	**5.57**	**6.57**
Mitsudo	100	63	0.59	5.67	9.59	15.85	16.85
Yamada	100	95	0.052	0.25	18.41	18.71	19.71

[a] Entry in bold text corresponds to present submission.
[b] AE = atom economy.
[c] PMI = process mass intensity.

(Figures 3.1 to 3.3 and Table 3.1) for CO_2 fixation of the propargyl amine to the oxazolidinone catalyzed by ion-exchange resin D301R under mild reaction conditions was developed (Scheme 3.2).[5]

REACTION PROCEDURES

SYNTHESIS OF N-BENZYLPROP-2-YN-1-AMINE FROM PROPARGYL BROMIDE[6]

A 100 mL two-neck flask containing a 20 mm magnetic stirring bar was charged with propargyl bromide (note 1, 1.64 mL, 80 wt.% in toluene [note 2, 0.46 g, 0.53 mL], 1.82 g, 15.30 mmol, 1.0 equiv), and benzylamine (note 3, 10.0 mL, 9.81 g, 91.6 mmol, 6.0 equiv) was added dropwise at 0°C to the reactor for 30 min. The reaction was stirred for 15 h at room temperature, and then 2 M NaOH (note 4,

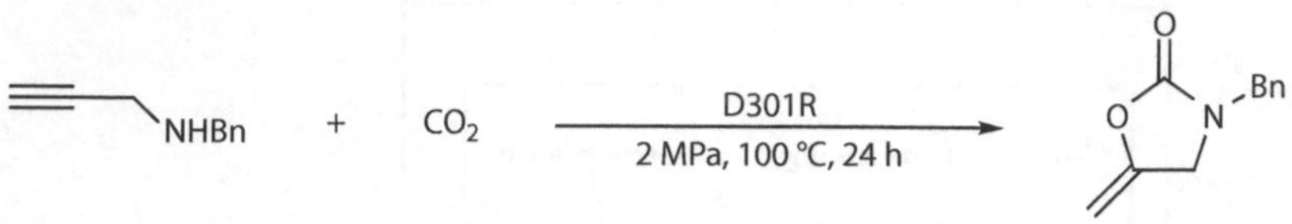

SCHEME 3.2 Synthesis of 3-benzyl-5-methyleneoxazolidin-2-one from N-benzylprop-2-yn-1-amine and CO_2.

1.2 g, distilled H_2O 15 mL) and Et_2O (20 mL, analytical reagent) were successively added. The organic layer was separated and the aqueous layer was extracted with Et_2O (2 × 10 mL). The combined organic layers were dried with $MgSO_4$ (≥99.0%, 1.5 g, analytical reagent) for 5 h, and the solvent was removed in vacuo (200 mmHg, 30°C) using a rotary evaporator. The residue was purified by flash chromatography on silica gel (note 5, 200–300 mesh, 16 g; eluting with 100:1 to 10:1 EtOAc (50 mL)/hexane (300 mL)) to afford the target compound as a colorless liquid (note 6, 1.654 g, 11.39 mmol, 75%).

The Procedure for the Cyclization Reactions of N-Benzylprop-2-yn-1-Amine with CO_2

In a 25 mL autoclave reactor with a quartzose tube and a 10 mm magnetic stirrer, the N-benzylprop-2-yn-1-amine (1 mmol, 145.2 mg), catalyst (note 7, D301R, 5 mol%, 0.05 mmol, 210.0 mg), and biphenyl (50 mg, an internal standard for gas chromatography (GC) analysis) were charged. Then CO_2 (note 8) was introduced into the autoclave. The pressure was adjusted to the designed pressure 2 MPa at 100°C, and the mixture was stirred for 24 h. After the reaction was completed (Scheme 3.3), the reactor was cooled in an ice-water bath and CO_2 was ejected slowly. An aliquot of sample was taken from the resultant mixture for GC analysis. The residue was purified by column chromatography on silica gel (note 9, 200–300 mesh, 10 g; eluting with 4:1 to 2:1 petroleum ether (note 10, 150 mL)/ethyl acetate (50 mL)). Pure product is obtained by distillation under reduced pressure (note 11, 169.7 mg, 89.7%).

NOTES

Instrumentation

[1]H nuclear magnetic resonance (NMR) spectra were recorded on a Bruker 300 or Bruker 400 spectrometer in $CDCl_3$, and tetramethylsilane (TMS, 0 ppm) was used as the internal reference. ^{13}C NMR was recorded at 75 or 100 MHz in $CDCl_3$, and $CDCl_3$ (77.0 ppm) was used as the internal reference. GC analysis was performed on a Shimadzu GC-2014, equipped with a capillary column (RTX-5, 30 × 0.25 m) using a flame ionization detector and electrospray ionization mass spectrometry (ESI-MS) with a spray voltage of 4.8 kV. High-resolution mass spectrometry was conducted using an Ionspec 7.0T spectrometer by electrospray ionization Fourier transform ion

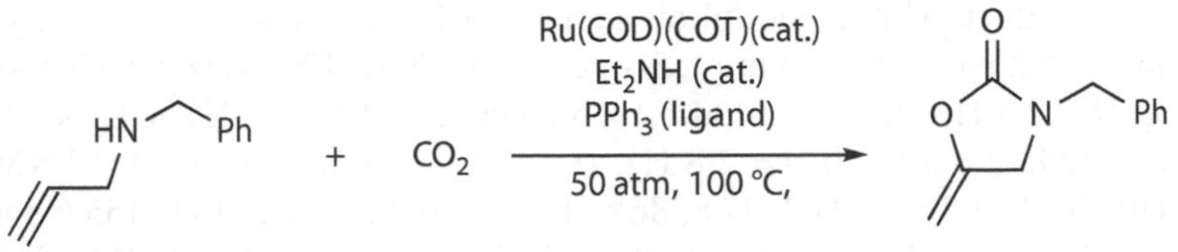

Ru(COD)(COT)(cat.)
Et$_2$NH (cat.)
PPh$_3$ (ligand)

HN⌐Ph   +   CO$_2$ $\longrightarrow$ 50 atm, 100 °C,

SCHEME 3.3 Ruthenium-catalyzed selective synthesis of enol carbamates by fixation of carbon dioxide.[7]

cyclotron resonance (ESI-FTICR) technique. Melting points were measured on an X$_4$ apparatus and uncorrected.

1. Propargyl bromide (>98%) was obtained from Yancheng Jiangsu China Chemical Co., Ltd. and used without further purification.
2. Toluene (≥99.5%, analytical reagent) was purchased from Tianjin Guangfu Fine Chemical Research Institute and used after being purified by standard techniques.
3. Benzylamine (≥99.0%, analytical reagent) was obtained from Kewei Tianjin University, China Chemical Co., Ltd. and used after being purified by standard techniques.
4. NaOH (colorless solid, ≥96.0%, analytical reagent) was obtained from Kewei Tianjin University, China Chemical Co., Ltd. The reagent was used as purchased.
5. Column was obtained from Beijing Synthware Glass Co., Ltd. (standard: inner diameter, 26 mm; effective length, 305 mm), and silica gel was purchased from Qingdao Haiyang Chemical, China. The eluant and silica gel were recycled by further purification.
6. N-Benzylprop-2-yn-1-amine: ^{1}H NMR (400 MHz, CDCl$_3$) δ = 1.71 (s, 1H, NH), 2.29 (d, $^3J_{HH}$ = 1.6 Hz, 1H, CH), 3.46 (s, 2H, CH$_2$), 3.92 (s, 2H, CH$_2$), 7.28–7.39 (m, 5H, Ar-H) ppm; ^{13}C NMR (100.6 MHz, CDCl$_3$) δ = 37.2, 52.1, 71.6, 81.9, 127.1, 128.4, 139.2 ppm. ESI-MS: Calculated for C$_{10}$H$_{11}$N 145.20, found 146.10 [M + H]$^+$.
7. The ion-exchange resins used in this study were commercially supplied by the Chemical Plant of Nankai University, Tianjin, China. D310R: Exchange capacity, 4.8 mmol g^{-1}; cross-linking degree of bead, 1%; true density, 1.03–1.07 g/mL; strength, ≥95%. Ion-exchange resins used in this study were evacuated at 373 K for 3 h.
8. Carbon dioxide with a purity of 99.99% was commercially available.
9. Column (standard: inner diameter, 17 mm; effective length, 305 mm) was obtained from Beijing Synthware Glass Co., Ltd., and silica gel was purchased from Qingdao Haiyang Chemical, China. The eluant and silica gel were recycled by further purification.
10. Petroleum ether (analytical reagent; boiling point range, 60–90°C) was used by further distillation under atmosphere condition.

11. The product exhibits the following physiochemical properties: light yellow solid; melting point (mp), 51–52°C; ^1H NMR (400 MHz, $CDCl_3$) 4.02 (t, $^3J = 2.4$ Hz, 2H), 4.24 (dd, $^3J = 5.2$ Hz, $^2J = 2.4$ Hz, 1H), 4.47 (s, 2H), 4.74 (dd, $^3J = 5.6$ Hz, $^3J = 2.8$ Hz, 1H), 7.29–7.39 (m, 5H) ppm; ^{13}C NMR (100 MHz, $CDCl_3$) 47.2, 47.8, 86.8, 128.2, 129.0, 134.9, 148.9, 155.6 ppm. ESI-MS: Calculated for $C_{11}H_{11}NO_2$ 189.21, found m/z 410.29 $(2M + Na)^+$.

SAFETY WARNING

Experiments using compressed CO_2 gases are potentially hazardous and must only be carried out using the appropriate equipment and under rigorous safety precautions.

IMPORTANCE ON GREEN CHEMISTRY CONTEXT

Data calculations: Offered by Dr. John Andraos.
Comparison with literature (Scheme 3.4):
He synthesis plan:

% Overall yield = 67
% Overall AE = 61
E-kernel = 1.36
E-excess = 8.45
E-aux = 24.96
E-total = 34.78
PMI = 35.78

Note: Metrics do not include chromatographic materials and preparation of catalysts.

Adrio–Carretero plan:[9]

% Overall yield = 79.7
% Overall AE = 40.4
E-kernel = 2.10
E-excess = 0.21
E-aux > 375.07
E-total > 377.39
PMI > 378.39

Note: Metrics do not include chromatographic materials and preparation of catalysts.

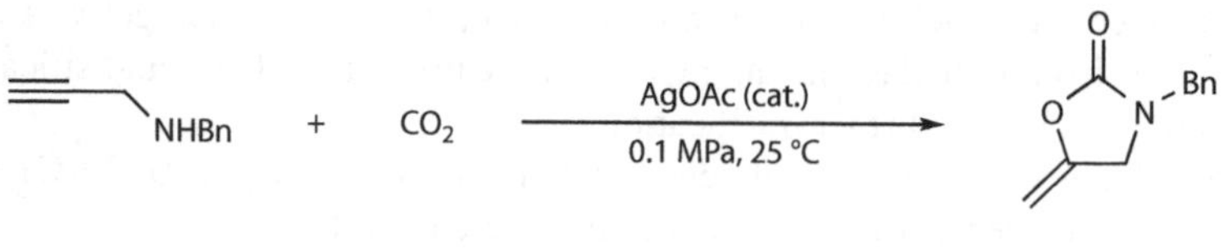

SCHEME 3.4 Silver-catalyzed carbon dioxide incorporation into propargylic derivative.[8]

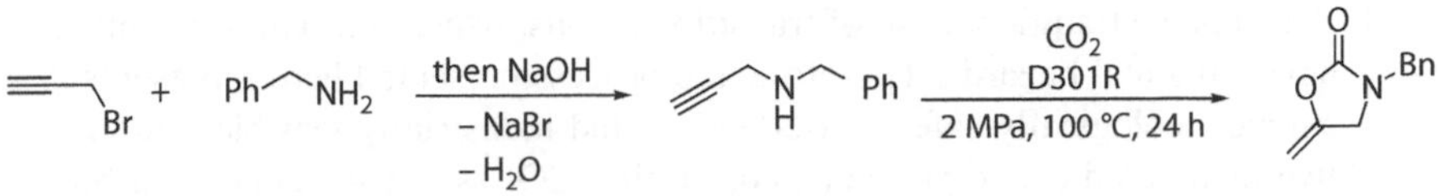

SCHEME 3.5 Procedures for the synthesis of 3-benzyl-5-methyleneoxazolidin-2-one.

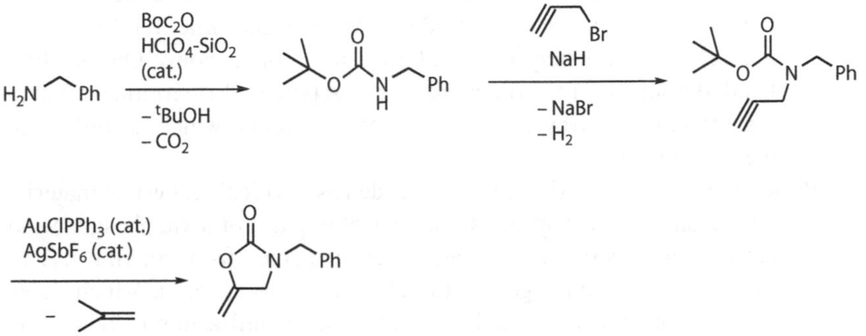

SCHEME 3.6 Gold-catalyzed synthesis of alkylidene 2-oxazolidinones.

Abbreviations and definitions:

AE = atom economy.
E-kernel = E-factor based on reaction by-products only.
E-excess = E-factor based on excess reagent consumption.
E-aux = E-factor based on all auxiliary materials used (reaction solvents, catalysts, workup materials, and purification materials).
E-total = total E-factor (sum of above three E-factors).
PMI = process mass intensity (total mass of input materials used divided by mass of target product collected as stated in reference 10).

The algorithms used to calculate metrics have been published in reference 11.

Conclusion: He's two-step plan (Scheme 3.5) has an overall process mass intensity that is at least 10 times less than the three-step Adrio–Carretero plan (Scheme 3.6).

EVALUATION FROM THE VIEWPOINT OF THE 12 PRINCIPLES

In this process, the insoluble ion-exchange resin D301R is demonstrated to be an effective catalyst for incorporation of CO_2. The present process has notable advantages and remarkable environmentally benign features:

1. It requires no additional organic solvents.
2. The catalyst is effective under mild conditions, and excellent yields together with good chemoselectivity could be attained.
3. The catalyst can be potentially recovered and kept its catalytic activity.[5]

4. It avoids the traditional use of transition metals, which may cause potential environmental hazards. Therefore, the process presented here represents a simple, ecologically safe, cost-effective, and industrially feasible route for five-membered cyclic products, consuming CO_2 as the sustainable carbon source. As a consequence, principles 5, 6, 7, and 9 have been highlighted in this procedure.

Principle 2: From this viewpoint, the selected reaction is a high atom-economic reaction, which maximizes the incorporation of all atoms of the substrate into the final product. At the same time, the high yield and selectivity are helpful to realizing the atom economy. On the other hand, the solvent-free process reduces the potential environmental hazard brought about by the waste organic solvents, which accords with the principle 5.

Principle 7: Carbon dioxide can be regarded as a typical renewable material and an alternative carbonylation reagent in place of toxic phosgene and carbon monooxide. The greatest benefit comes from making use of the main greenhouse gas as the chemical raw material. which eases the environment pressure. In this context, the utilization of renewable feedstock, a less toxic reagent, with a simple procedure would be the important feature and significant to green organic synthesis.

REFERENCES

1. (a) T. Sakakura, J.-C. Choi, H. Yasuda, *Chem. Rev.*, 2007, 107, 2365–2387; (b) T. Sakakura, K. Kohno, *Chem. Commun.*, 2009, 1312–1330.
2. For recent reviews on transformation of carbon dioxide, see: (a) X.-B. Lu, D. J. Darensbourg, *Chem. Soc. Rev.*, 2012, 41, 1462–1484; (b) I. Omae, *Coord. Chem. Rev.*, 2012, 256, 1384–1405; (c) M. Cokoja, C. Bruckmeier, B. Rieger, W. A. Herrmann, F. E. Kühn, *Angew. Chem. Int. Ed.*, 2011, 50, 8510–8537; (d) K. M. K. Yu, I. Curcic, J. Gabriel, S. C. E. Tsang, *ChemSusChem*, 2008, 1, 893–899; (e) D. B. Dell'Amico, F. Calderazzo, L. Labella, F. Marchetti, G. Pampaloni, *Chem. Rev.*, 2003, 103, 3857–3898; (f) I. Omae, *Catal. Today*, 2006, 115, 33–52; (g) G. W. Coates, D. R. Moore, *Angew. Chem. Int. Ed.*, 2004, 43, 6618–6639; (h) D. J. Darensbourg, M. W. Holtcamp, *Coord. Chem. Rev.*, 2003, 103, 3857–3897; (i) X. L. Yin, J. R. Moss, *Coord. Chem. Rev.*, 1999, 181, 27–59.
3. Application of oxazolidinones as intermediates: (a) R. J. Watson, D. Batty, A. D. Baxter, D. R. Hannah, D. A. Owen, J. G. Montana, *Tetrahedron Lett.*, 2002, 43, 683–685; (b) L. Aurelio, R. T. C. Brownlee, A. B. Hughus, *Chem. Rev.*, 2004, 104, 5823–5846; (c) T. M. Makhtar, G. D. Wright, *Chem. Rev.*, 2005, 105, 529–542; (d) T. Andreou, A. M. Costa, L. Esteban, L. Gonzalez, G. Mas, J. Vilarrasa, *Org. Lett.*, 2005, 7, 4083–4086.
4. Application of oxazolidinones as chiral auxiliaries: (a) C. W. Phoon, C. Abell, *Tetrahedron Lett.*, 1998, 39, 2655–2658; (b) M. Prashad, Y. G. Liu, H. Y. Kim, O. Repic, T. J. Blacklock, *Tetrahedron Asymm.*, 1999, 10, 3479–3482; (c) R. E. Gawley, S. A. Campagna, M. Santiago, T. Ren, *Tetrahedron Asymm.*, 2002, 13, 29–36.
5. A.-H. Liu, L.-N. He, S.-Y. Peng, Z.-D. Pan, J.-L. Wang, J. Gao, *Sci. China Chem.*, 2010, 53, 1578–1585.
6. (a) W. Hess, J. W. Burton, *Chem.-Eur. J.*, 2010, 16, 12303–12306; (b) C. Molinaro, T. F. Jamison, *J. Am. Chem. Soc.*, 2003, 125, 8076–8077; (c) J. Gao, Q.-W. Song, L.-N. He, C. Liu, Z.-Z. Yang, X. Han, X.-D. Li, Q.-C. Song, *Tetrahedron*, 2012, 68, 3835–3842.

7. T. Mitsudo, Y. Hori, Y. Yamakawa, Y. Watanabe, *Tetrahedron Lett.*, 1987, 28, 4417–4418.
8. S. Kikuchi, S. Yoshida, Y. Sugawara, W. Yamada, H. M. Cheng, K. Fukui, K. Sekine, I. Iwakura, T. Ikeno, T. Yamada, *Bull. Chem. Soc. Jpn.*, 2011, 84, 698–717.
9. (a) R. Robles-Machin, J. Adrio, J. C. Carretero, *J. Org. Chem.*, 2006, 71, 5023–5026; (b) A. K. Chakraborti, S. V. Chankeshwara, *Org. Biomol. Chem.*, 2006, 4, 2769–2771.
10. C. Jiménez-González, C. S. Ponder, Q. B. Broxterman, J. B. Manley, *Org. Proc. Res. Dev.*, 2011, 15, 912–917.
11. (a) J. Andraos, *Org. Process Res. Dev.*, 2009, 13, 161–185; (b) J. Andraos, M. Sayed, *J. Chem. Educ.*, 2007, 84, 1004–1011; (c) J. Andraos, *The Algebra of Organic Synthesis: Green Metrics, Design Strategy, Route Selection, and Optimization*, CRC Press, Boca Raton, FL, 2012.

4 Reaction 2: Synthesis of the Five-Membered Cyclic Carbonates from Epoxides and CO$_2$

Qing-Wen Song and Liang-Nian He

CONTENTS

INTRODUCTION

Recently, CO$_2$ fixation has received much attention because CO$_2$ is the most inexpensive and renewable carbon resource from the viewpoint of green chemistry.[1] The development of a truly environmentally friendly process utilizing CO$_2$, which is the largest single source of greenhouse gas and can be also regarded as a typical renewable natural resource, has drawn current interest in organic synthetic chemistry from the viewpoint of environmental protection and resource utilization. Chemical fixation of CO$_2$ onto industrial useful materials is one of the most promising methods because there are many possibilities for CO$_2$ to be used as a safe and cheap C$_1$ building block in organic synthesis.[2] Organic cyclic carbonates such as ethylene carbonate and propylene carbonate are widely used for various purposes, for instance, electrolytic elements of lithium secondary batteries, polar aprotic solvents, monomers for synthesizing polycarbonates, and chemical ingredients for preparing medicines or agricultural chemicals.[3]

Polyethylene glycol (PEG) and its derivatives are known to be inexpensive, thermally stable, recoverable, toxicologically innocuous, and environmentally benign media[4] for chemical reactions, and have almost negligible vapor pressure and

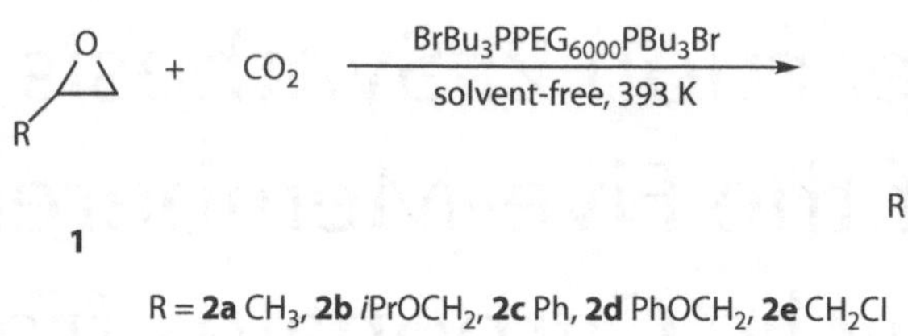

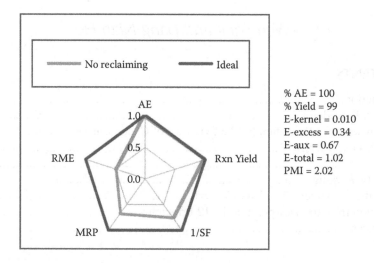

$R = \mathbf{2a}\ CH_3,\ \mathbf{2b}\ iPrOCH_2,\ \mathbf{2c}\ Ph,\ \mathbf{2d}\ PhOCH_2,\ \mathbf{2e}\ CH_2Cl$

SCHEME 4.1 Synthesis of various carbonates catalyzed by phosphonium halide-functionalized polyethylene glycol.

FIGURE 4.1 Radial pentagon for synthesis of 4-methyl-[1,3]-dioxolan-2-one (this work). (Note: Metrics do not include preparation of catalysts.)

biphasic catalysis in supercritical CO_2. The properties of PEG offer the possibility for immobilizing and recycling a homogeneous catalyst.[4c,d,f] PEG derivatives were covalently bound to a quaternary ammonium salt as a homogeneous catalyst for CO_2 fixation in quantitative yield and excellent selectivity, also leading to the catalyst recycle (Scheme 4.1). The process could be efficient and environmentally benign (see Figure 4.1), and have simple operation, facile separation of the product, and easy recycles of the catalyst.

REACTION PROCEDURES

REPRESENTATIVE PROCEDURE FOR THE SYNTHESIS OF CYCLIC CARBONATE CATALYZED BY $PEG_{6000}(PBU_3BR)_2$[5]

The reaction was carried out in a stainless steel autoclave reactor with an inner volume of 25 mL. In the autoclave equipped with a 10 mm magnetic stirrer, $PEG_{6000}(PBu_3Br)_2$[6] (note 1, 1 mol%, 1.86 g) was added to a solution of propylene oxide

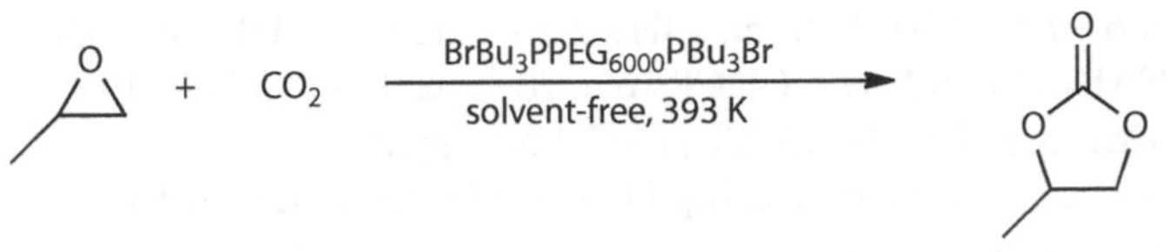

SCHEME 4.2 Synthesis of propylene carbonate from propylene oxide and CO_2.

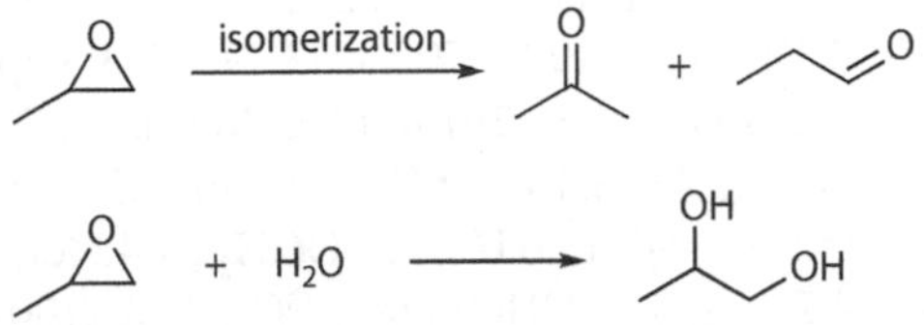

SCHEME 4.3 Side reaction detected in the process of propylene carbonate synthesis.

(note 2, 28.60 mmol, 1.66 g, 2.0 mL) and biphenyl (80 mg, as internal standard for GC) under nitrogen. CO_2 was then introduced into the autoclave with an initial pressure of ca. 5.0 MPa at room temperature (Scheme 4.2). The pressure was generally adjusted to 1 MPa at 393 K. The mixture was stirred for 6 h, and the temperature was kept constant during the reaction. At the end of the reaction, the autoclave was cooled to ambient temperature and further cooled to ca. 0°C with an ice-water bath. The excess of carbon dioxide was ejected slowly. The products were detected by gas chromatograph equipped with a flame ionization detector and a capillary column (Agilent technologies 6890, HP-5. 0.25 mm i.d. × 30 m), and further determined by GC-MS (HP G1800A) by comparing retention times and fragmentation patterns with authentic samples (Scheme 4.3). All carbonates (note 3, **2a–2e**) are known compounds, which were identified by ${}^1H$ and ${}^{13}C$ nuclear magnetic resonance (NMR) spectra and GC-MS.[7]

NOTES

The chemical reagents and solvents were used as received without further purification unless otherwise mentioned according to the references. After reaction, purification of $PEG_{6000}(NBu_3Br)_2$ involved evaporation of the reaction solvent and addition of the residue dissolved in CH_2Cl_2 to a stirred solution of Et_2O cooled at 0°C. After 0.5 h, the obtained suspension was filtered and the solid was washed with Et_2O. All of NMR spectra were recorded with a Varian Mercury-Plus 400 spectrometer at 400 MHz (${}^1H$ NMR) and 100.6 MHz (${}^{13}C$ NMR) respectively. Chemical shifts were referenced to internal solvent resonances.

1. The preparing process of catalyst was described in reference 6 and dried at 60°C under the vacuum condition for 24 h before used. Characteristic data of $BrBu_3PPEG_{6000}PBu_3Br$: ${}^1H$ NMR (400 MHz, $CDCl_3$) δ = 0.96

(t, J = 6.8 Hz, 18H), 1.50 (m, 24H), 2.36–2.42 (m, 12H), 3.44–3.81 (m) ppm. ^{13}C NMR (100.6 MHz, CDCl$_3$) δ = 70.9, 62.0, 30.5, 24.0, 19.8, 13.7 ppm. ^{31}P NMR (161.7 MHz, CDCl$_3$) δ = 34.5 (s) ppm.

2. Propylene oxide (colorless liquid, ≥99.5%) was obtained from Alfa Aesar and used as purchased.

3. Detailed NMR data information.[8]

4-Methyl-1,3-dioxolan-2-one (**2a**) (2.89 g, 99%): ^{1}H NMR (400 MHz, CDCl$_3$) δ = 1.43 (d, J = 6.0 Hz, 3H, Me), 3.98 (t, J = 8.4 Hz, 1H, OCH$_2$), 4.51 (t, J = 8.4 Hz, 1H, OCH$_2$), 4.82 (m, 1H, CHO) ppm; ^{13}C NMR (100.4 MHz, CDCl$_3$) δ = 19.15, 70.53, 73.49, 154.95 ppm.

Isopropoxy-1,3-dioxolan-2-one (**2b**) (4.12 g, 90%): ^{1}H NMR (400 MHz, CDCl$_3$) δ = 1.08 (t, J = 6.4 Hz, 6H, 2 × Me), 3.51–3.62 (m, 3H, CHO, CH$_2$O), 4.30 (dd, J = 8.0, 15.6 Hz, 1H, OCH$_2$), 4.42 (dd, J = 8.0, 15.6 Hz, 1H, OCH$_2$), 4.74 (m, 1H, CHO) ppm; ^{13}C NMR (100.4 MHz, CDCl$_3$) δ = 21.53, 21.65, 66.16, 66.89, 72.59 ppm.

4-Phenyl-1,3-dioxolan-2-one (**2c**) (5.50 g, 99%): ^{1}H NMR (400 MHz, CDCl$_3$) δ = 4.35 (t, J = 8.4 Hz, 1H, OCH$_2$), 4.80 (t, J = 8.4 Hz, 1H, OCH$_2$), 5.70 (t, J = 8.0 Hz, 1H, OCH), 7.36 (d, J = 7.6 Hz, 2H, Ph), 7.44 (d, J = 6.4 Hz, 3H, Ph) ppm; ^{13}C NMR (100.4 MHz, CDCl$_3$) δ = 71.12, 77.95, 125.83, 129.22, 129.71, 135.78, 154.76 ppm.

4-Phenoxymethyl-1,3-dioxolan-2-one (**2d**) (3.63 g, 93%): ^{1}H NMR (400 MHz, CDCl$_3$) δ = 4.15 (dd, J = 4.4, 10.8 Hz, 1H, OCH$_2$), 4.24 (dd, J = 3.6, 10.8 Hz, 1H, OCH$_2$), 4.55 (dd, J = 6.0, 8.4 Hz, 1H, PhOCH$_2$), 4.62 (t, J = 8.4 Hz, 1H, PhOCH$_2$), 5.03 (m, 1H, OCH), 6.91 (d, J = 8.0 Hz, 2H, Ph), 7.02 (t, J = 7.4 Hz, 1H, Ph), 7.31 (t, J = 8.0 Hz, 2H, Ph) ppm; ^{13}C NMR (100.4 MHz, CDCl$_3$) δ = 66.17, 68.84, 74.11, 114.57, 121.92, 129.63, 154.65, 157.71 ppm.

4-Chloromethyl-1,3-dioxolan-2-one (**2e**) (4.18 g, 89%): ^{1}H NMR (400 MHz, CDCl$_3$) δ = 3.71 (dd, J = 3.2, 12.0 Hz, 1H, ClCH$_2$), 3.80 (dd, J = 5.2, 12.0 Hz, 1H, ClCH$_2$), 4.39 (dd, J = 6.0, 8.4 Hz, 1H, OCH$_2$), 4.58 (t, J = 8.4 Hz, 1H, OCH$_2$), 4.98 (m, 1H, CHO) ppm; ^{13}C NMR (100.4 MHz, CDCl$_3$) δ = 43.84, 66.83, 74.29, 154.28 ppm.

SAFETY WARNING

Experiments using compressed CO$_2$ gases are potentially hazardous and must only be carried out using the appropriate equipment and under rigorous safety precautions.

IMPORTANCE ON GREEN CHEMISTRY CONTEXT

Data calculations: Offered by Dr. John Andraos.

Abbreviations and definitions pertaining to Tables 4.1 and 4.2:

AE = atom economy.
E-kernel = E-factor based on reaction by-products only.
E-excess = E-factor based on excess reagent consumption.

TABLE 4.1

Green Metrics Summary for Synthesis of 4-Methyl-[1,3]-Dioxolan-2-One from Propylene Oxide[a,b]

Catalyst	% AE	% Yield	E-Kernel	E-Excess	E-Aux	E-Total	PMI
nBu$_4$NBr/SiO$_2$[9]	100	96	0.041	0.0034	0.27	0.31	1.31
[H-DBU]Cl[10]	100	97	0.031	0.22	0.089	0.34	1.34
PEG$_{6000}$(NBu$_3$Br)$_2$[6a]	100	98	0.020	0.50	0.031	0.55	1.55
PEG$_{2000}$(PBu$_3$Br)$_2$[11]	100	99	0.0094	0.50	0.28	0.79	1.79
CrOTPP/1-methyl-imidazole[15]	100	90	0.11	0.72	0.0019	0.82	1.82
PEG$_{6000}$(PBu$_3$Br)$_2$	**100**	**99**	**0.010**	**0.34**	**0.67**	**1.02**	**2.02**
Chitosan-ZnCl$_2$[12]	100	97	0.031	3.07	0.058	3.16	4.16
ZnCl$_2$/supported ionic liquid[13]	100	95	0.054	3.04	0.44	3.53	4.53
Pyridinium ionic liquid[14]	100	99	0.0099	1.42	20.14	21.57	22.57

[a] Metrics do not include preparation of catalysts; CrOTPP = chromium (IV) oxide tetra-*p*-tolylporphyrinate; DBU = 1,8-diazabicyclo[5.4.0]undec-7-ene; PEG = polyethylene glycol.

[b] Entry in bold text corresponds to present submission.

TABLE 4.2

Green Metrics Summary for Synthesis of 4-Methyl-[1,3]-Dioxolan-2-One from Propane-1,2-Diol[a]

Conditions	% AE	% Yield	E-Kernel	E-Excess	E-Aux	E-Total	PMI
Bu$_2$Sn = O[16]	85	1.7	67.28	0.19	18.35	85.81	86.81
Mg[17]	70.7	2.8	49.90	64,675	571	65,296	65,297

[a] Metrics do not include preparation of catalysts. For detailed reaction conditions, see Scheme 4.4.

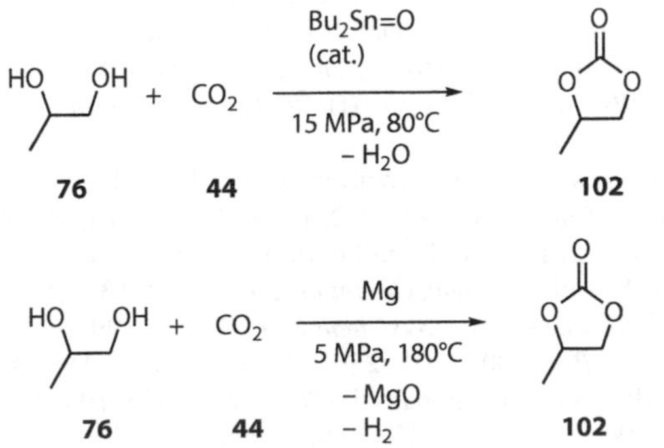

SCHEME 4.4 Synthesis of ester carbonate from 1,2-propylene glycol and CO$_2$.

E-aux = E-factor based on all auxiliary materials used (reaction solvents, catalysts, workup materials, and purification materials).

E-total = total E-factor (sum of above three E-factors).

PMI = process mass intensity (total mass of input materials used divided by mass of target product collected as stated in reference 18).

The algorithms used to calculate metrics have been published in reference 19.

Conclusion: He's catalytic strategy has an environment-friendly process (see Figure 4.1 and Scheme 4.1) in comprehensive consideration compared with other catalytic systems listed in publications.

EVALUATION FROM THE VIEWPOINT OF THE 12 PRINCIPLES

The catalytic process presented here represents a simple, ecologically safe, and cost-effective route to five-membered cyclic products in an almost quantitative yield and selectivity without utilizing organic solvent, consuming CO_2 as the sustainable carbon source. This procedure would feature one-step operation under relatively mild conditions for the production of cyclic carbonate marked as a chemical product with low toxicity by using recyclable homogeneous catalyst. This is a perfect atom economical reaction. In particular, PEG and its derivatives could be inexpensive, thermally stable, recoverable, toxicologically innocuous, and environmentally benign solvents and catalysts for chemical reactions, and have almost negligible vapor pressure. PEG derivative can be covalently bound to a quaternary ammonium salt as a homogeneous catalyst for CO_2 fixation with good results as well as exceptional catalyst recovery.

As a consequence, principles 2, 4, 5, 7, 9, and 10 have been highlighted in this procedure.

REFERENCES

1. For recent reviews on fixation of carbon dioxide, see: (a) D. J. Darensbourg, R. M. Mackiewicz, A. L. Phelps, D. R. Billodeaux, *Acc. Chem. Res.* 2004, 37, 836–844; (b) P. Braunstein, D. Matt, D. Nobel, *Chem. Rev.* 1988, 88, 747–764; (c) K. C. Nicolaou, Z. Yang, J. J. Liu, H. Ueno, P. G. Nantermet, R. K. Guy, C. F. Claiborne, J. Renaud, E. A. Couladouros, K. Paulvannanand, E. J. Sorensen, *Nature* 1994, 367, 630–634; (d) P. G. Jessop, T. Ikariya, R. Noyori, *Chem. Rev.* 1995, 95, 259–272; (e) D. H. Gibson, *Chem. Rev.* 1996, 96, 2063–2095; (f) W. Leitner, *Coord. Chem. Rev.* 1996, 153, 257–284.
2. For recent reviews on transformation of carbon dioxide, see: (a) X.-B. Lu, D. J. Darensbourg, *Chem. Soc. Rev.* 2012, 41, 1462–1484; (b) M. Cokoja, C. Bruckmeier, B. Rieger, W. A. Herrmann, F. E. Kühn, *Angew. Chem. Int. Ed.* 2011, 50, 8510–8537; (c) T. Sakakura, K. Kohno, *Chem. Commun.* 2009, 1312–1330; (d) K. M. K. Yu, I. Curcic, J. Gabriel, S. C. E. Tsang, *ChemSusChem* 2008, 1, 893–899; (e) T. Sakakura, J. C. Choi, H. Yasuda, *Chem. Rev.* 2007, 107, 2365–2387; (f) D. B. Dell'Amico, F. Calderazzo, L. Labella, F. Marchetti, G. Pampaloni, *Chem. Rev.* 2003, 103, 3857–3898; (g) I. Omae, *Catal. Today* 2006, 115, 33–52; (h) G. W. Coates, D. R. Moore, *Angew. Chem. Int. Ed.* 2004, 43, 6618–6639; (i) D. J. Darensbourg, M. W. Holtcamp, *Coord. Chem. Rev.* 2003, 103, 3857–3897; (j) X. L. Yin, J. R. Moss, *Coord. Chem. Rev.* 1999, 181, 27–59.

3. (a) A. A. G. Shaikh, S. Sivaram, *Chem. Rev.* 1996, 96, 951–976; (b) M. Yoshida, M. Ihara, *Chem. Eur. J.* 2004, 10, 2886–2893; (c) B. Schäffner, F. Schäffner, S. P. Verevkin, A. Börner, *Chem. Rev.* 2010, 110, 4554–4581.

4. For typical references on PEG as benign media, see: (a) D. J. Heldebrant, P. G. Jessop, *J. Am. Chem. Soc.* 2003, 125, 5600–5601; (b) R. Annunziata, M. Benaglia, M. Cinquini, F. 0Cozzi, G. Tocco, *Org. Lett.* 2000, 2, 1737–1739; (c) M. T. Reetz, W. Wiesenhöfer, *Chem. Commun.* 2004, 2750–2751; (d) J. Chen, S. K. Spear, J. G. Huddleston, R. D. Rogers, *Green Chem.* 2005, 7, 64–82, and references cited therein; (e) S. Chandrasekhar, Ch. Narsihmulu, S. S. Sultana, N. R. Reddy, *Chem. Commun.* 2003, 1716–1717; (f) M. Solinas, J. Jiang, O. Stelzer, W. Leitner, *Angew. Chem. Int. Ed.* 2005, 44, 2291–2295; (g) Z. Hou, N. Theyssen, A. Brinkmann, W. Leitner, *Angew. Chem. Int. Ed.* 2005, 44, 1346–1349.

5. J.-S. Tian, C.-X. Miao, J.-Q. Wang, F. Cai, Y. Du, Y. Zhao, L.-N. He, *Green Chem.* 2007, 9, 566–571.

6. (a) Y. Du, J.-Q. Wang, J.-Y. Chen, F. Cai, J.-S. Tian, D.-L. Kong, L.-N. He, *Tetrahedron Lett.* 2006, 47, 1271–1275; (b) L.-N. He, J.-S. Tian, China patent, application 2006100149098, 2006; (c) S. Grinberg, E. Shaubi, *Tetrahedron* 1991, 47, 2895–2902; (d) R. Annunziata, M. Benaglia, M. Cinquini, F. Cozzi, G. Tocco, *Org. Lett.* 2000, 2, 1737–1739.

7. Possible side reactions: The products other than propylene carbonate were isomers of propylene oxide, such as acetone, propionaldehyde, 1,2-propanediol, and so on.

8. X.-Y. Dou, J.-Q. Wang, Y. Du, E. Wang, L.-N. He, *Synlett* 2007, 3058–3062.

9. J.-Q. Wang, D.-L. Kong, J.-Y. Chen, F. Cai, L.-N. He, *J. Mol. Catal. A* 2006, 249, 143–148.

10. Z.-Z. Yang, L.-N. He, C.-X. Miao, S. Chanfreau, *Adv. Synth. Catal.* 2010, 352, 2233–2240.

11. J.-S. Tian, F. Cai, J.-Q. Wang, Y. Du, L.-N. He, *Phosphorus Sulfur Silicon* 2008, 183, 494–498.

12. L.-F. Xiao, F.-W. Li, C.-G. Xia, *Appl. Catal. A* 2005, 279, 125–129.

13. L.-F. Xiao, F.-W. Li, J.-J. Peng, C.-G. Xia, *J. Mol. Catal. A* 2006, 253, 265–269.

14. W.-L. Wong, P.-H. Chan, Z.-Y. Zhou, K.-H. Lee, K.-C. Cheung, K.-Y. Wong, *ChemSusChem* 2008, 1, 67–70.

15. W. J. Kruper, D. V. Dellar, *J. Org. Chem.* 1995, 60, 725–727.

16. Y. Du, D.-L. Kong, H.-Y. Wang, F. Cai, J.-S. Tian, J.-Q. Wang, L.-N. He, *J. Mol. Catal. A* 2005, 241, 233–237.

17. Y. Du, L.-N. He, D.-L. Kong, *Catal. Commun.* 2008, 9, 1754–1758.

18. C. Jiménez-González, C. S. Ponder, Q. B. Broxterman, J. B. Manley, *Org. Proc. Res. Dev.* 2011, 15, 912–917.

19. (a) J. Andraos, *Org. Process Res. Dev.* 2009, 13, 161–185; (b) J. Andraos, M. Sayed, *J. Chem. Educ.* 2007, 84, 1004–1011; (c) J. Andraos, *The Algebra of Organic Synthesis: Green Metrics, Design Strategy, Route Selection, and Optimization*, CRC Press, Boca Raton, FL, 2012.

5 Green Methods for the Epoxidation of Olefins Part I: Using Venturello Anion-Based Catalysts

Yunxiang Qiao and Zhenshan Hou

CONTENTS

The catalytic epoxidation of olefins constitutes an important research area and represents the core of a variety of chemical processes for producing bulk and fine chemicals. There are many name reactions for epoxidation, e.g., Jacobsen–Katsuki epoxidation, Prilezhaev reaction, Sharpless epoxidation, Shi epoxidation, etc., as well as other processes using various catalysts, oxygen sources, reaction media, and additives. The conventional processes for olefin epoxidation are still the chlorohydrin process (a chlorine-using noncatalytic process) and catalytic processes where organic peroxides are used extensively. The disadvantages of these classical processes are quite obvious from the economic and environmental viewpoints.[1]

On the other hand, a polyoxometalate (POM) is a polyatomic anion, that consists of three or more transitionmetal oxyanions linked together by shared oxygen atoms to form a large, closed three-dimensional framework. The metal atoms are usually group 5 or group 6 transition metals in their high oxidation states. In this state, their electron configuration is d^0 or d^1. Examples include vanadium(V), niobium(V), tantalum(V), molybdenum(VI), and tungsten(VI). The framework of transition metal oxyanions may enclose one or more heteroatoms, such as phosphorus or silicon, themselves sharing neighboring oxygen atoms with the framework. There are basically six different POM structure types: Keggin, Dawson, Anderson, Linqvist, Wangh, and Silverton structures. Generally, Keggin-type heteropoly compounds (e.g., $H_3PW_{12}O_{40}$) are the most stable ones among various POMs.

Due to the versatility, accessibility, and redox properties, the transition metal peroxo complexes are widely used in olefin epoxidation reactions[1-5] by using different oxygen sources, reaction media, additives, etc. There are a great number of results reported on olefin epoxidation with POMs as catalysts, exemplified by classical Venturello's anion ($\{PO_4[WO(O_2)_2]_4\}^{3-}$),[6,7] Ishii's anion ($[PW_{12}O_{40}]^{3-}$),[8] Mizuno's anions ($[W_2O_3(O_2)_4]^{2-}$ and $[\gamma\text{-}SiW_{10}O_{34}]^{4-}$),[9-12] Neumann's anion ($[WZnMn_2 (ZnW_9O_{34})_2]^{12-}$),[13] Xi's anion ($[PW_4O_{16}]^{3-}$),[14] etc. This chapter mainly discusses the greening process for epoxidation using phosphotungstate-based self-separated catalysts that were developed in our laboratory or elsewhere. Unless otherwise indicated, all the reactions were performed with H_2O_2 as an oxidant under air atmosphere.

EPOXIDATION USING THE CLASSICAL VENTURELLO–ISHII CATALYTIC SYSTEM

In 1983, Venturello et al. proposed that the complex consisting of tungstate and phosphate under phase-transfer conditions can catalyze the epoxidation of different alkenes with dilute H_2O_2 solution (10%) as oxidant and dichloroethane or benzene as the organic phase (Scheme 5.1).[6]

The catalytic species is formed in situ in the reaction mixture by simply introducing water-soluble alkaline tungstates and phosphoric (or arsenic) acid or its alkaline salts in a 1:2 molar ratio. Otherwise, only poor or no catalytic activity can be obtained. High selectivities to epoxide (80–90%) and both hydrogen peroxide and olefin at a substantially complete conversion are usually attained after relatively short reaction times and under mild conditions.

The effectiveness of the method depends largely on the pH of the aqueous phase. It increases as the pH decreases, and to the extent permitted by stability to hydrolysis of the epoxide formed, low pH values are preferable. For epoxidation of long-chain olefins such as 1-octene, 1-dodecene, and the like, pH = 1.6 is satisfactory. In spite of the acidic conditions, hydrolytic cleavage of the epoxy ring in general is largely prevented. High selectivities can be attributed to both the protecting effect of the double phase and the relatively short contact times

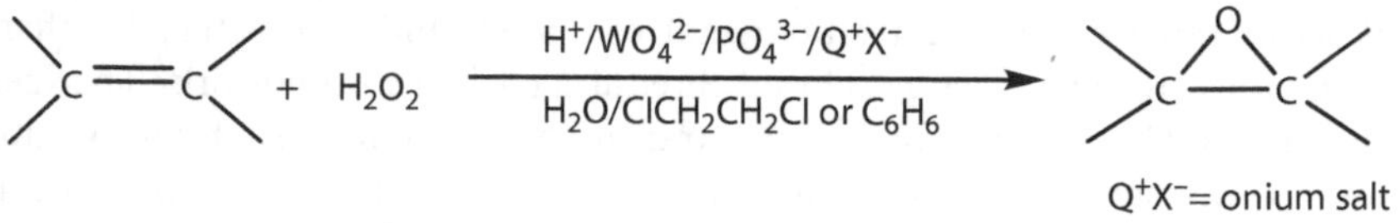

$Q^+X^- =$ onium salt

SCHEME 5.1 Epoxidation of various alkenes under phase-transfer conditions in the Venturello system.

with the aqueous phase, which are made possible by the high efficiency of the catalyst and by the use of an excess of olefin. When the olefin was employed in a stoichiometric amount with respect to hydrogen peroxide, a substantial reduction (30–80%) in yield was observed. It should be noted that also gaseous olefins such as propylene and butadiene could be epoxidized. In these cases, however, yields were considerably lower (30–40%) than those obtained with the other substrates tested. Butadiene gave the monoepoxide exclusively. The reaction appears to be stereospecific, since only *trans*-2-hexene oxide was detected by either gas-liquid chromatography (GLC) or ^{1}H nuclear magnetic resonance (NMR) analysis when *trans*-2-hexene was subjected to epoxidation. Analogously, *cis*-2-hexene gave only *cis*-2-hexene oxide.

Chlorinated and aromatic hydrocarbons were chosen as solvents, and phase-transfer agents (Q^+X^-) like methyltrioctylammonium chloride (Aliquat 336), dimethyl[dioctadecyl (75%) + dihexadecyl (25%)]ammonium chloride (Arquad 2HT), and hexadecyltributylphosphonium chloride can lead to successful oxidation. The reaction may also be run without solvent when the olefin is a liquid, with only slightly diminished yields.

In a typical experiment, finely powdered $Na_2WO_4 \cdot 2H_2O$ (1.65 g, 5 mmol), 40% w/v H_3PO_4 (2.45 mL, 10 mmol), and 8% aqueous H_2O_2 (51 g, 120 mmol) were introduced into a glass reactor. The pH of the aqueous solution was adjusted to 1.6 by 30% H_2SO_4; thereupon 1-octene (22.4 g, 200 mmol), 1,2-dichloroethane (15 mL), and the phase-transfer agent—methyltrioctylammonium chloride (abbreviated as MTOC, 0.82 g, ca. 2 mmol)—were added. Under vigorous stirring, the resultant biphase mixture was heated to 70°C for 45 min. The water and 1,2-chloroethane layers were then separated, and then the amount of unreacted H_2O_2 was determined by iodometric titration of the aqueous phase.[15] The organic layer was analyzed by GLC using an internal standard (β,β′-oxydipropionitrile column). In this way, H_2O_2 conversion was found to be 98%, and 12.7 g (99 mmol, in 82.5% of yield based on H_2O_2 charged) of 1,2-epoxyoctane was obtained. However, the yield of epoxides from propylene and butadiene was considerably lower (30–40%) than that from higher terminal olefins.

In 1988, Ishii et al. proposed that the system composed of $H_3PW_{12}O_{40}$ and cetylpyridinium chloride (CPC) ($H_3PW_{12}O_{40}$) can catalyze epoxidation of alkenes with commercially available H_2O_2 solution as oxidant and benzene or t-butyl alcohol as organic phase.[8] In a typical experiment, to a stirred solution of $H_3PW_{12}O_{40}$ (74.9 mg, ca. 0.026 mmol), CPC (28 mg, 0.082 mmol) and 35% H_2O_2 (960 mg, 9.8 mmol) in chloroform (15 mL) were added to 1-octene (728 mg, 6.5 mmol), and the mixture was allowed to react at 60°C for 5 h. After the partially precipitated catalyst was separated by filtration, the filtrate was treated with a solution of 10% sodium hydrogen sulfite to decompose unreacted H_2O_2 and then with 10% sodium hydroxide. The product was extracted with chloroform. Pure product 1,2-epoxyoctane (yield 80.4%) was obtained by distillation or silica gel column chromatography (hexane/ethyl acetate, 10/1–2, eluent). Spectral data of the product were compared with those of authentic sample and reported in the literature.

PEROXOPOLYOXOMETALATE-BASED ROOM TEMPERATURE IONIC LIQUID AS A SELF-SEPARATION CATALYST FOR EPOXIDATION OF OLEFINS

A new peroxopolyoxometalate-based room temperature ionic liquid (POM-RTIL) was synthesized and used as catalyst for efficient epoxidation of various olefins.[16] It was the first example of olefin epoxidation catalyzed by a POM-based RTIL. The RTIL catalyst was found to be well dissolved in the solvent (ethyl acetate) during the reaction, while it can self-separate from the reaction media at room temperature after the reaction was completed, which made the recovery and reuse of the IL catalyst convenient. The POM-RTIL catalyst can be recycled five times without significant loss of activity under the condition of no solvent or using ethyl acetate as solvent.

The catalyst POM-RTIL is synthesized in three steps, as shown in Scheme 5.2.

Step 1: Chlorination of PEG-300. $SOCl_2$ (2.4 g, 20 mmol) in benzene (10 mL) was slowly added to a mixture of dried PEG-300 (1.8 g, 6 mmol), pyridine (1.6 g, 20 mmol), and benzene (15 mL) over 0.5 h at room temperature. The mixture was heated for 8 h at reflux under nitrogen and cooled to room temperature. Then the solution was diluted with CH_2Cl_2 (20 mL) and the mixture was filtrated to remove precipitate. The organic phase was washed with saturated NaCl solution (15 mL × 3) and distilled water (15 mL × 2), and then dried at 50°C for 3 h under vacuum to afford the product (1.7 g, 85% yield).

Step 2: PEG chain-functionalized N-dodecylimidazolium dichlorides. Chlorinated PEG-300 (505 mg, 1.5 mmol) and dodecylimidazole[17–19]

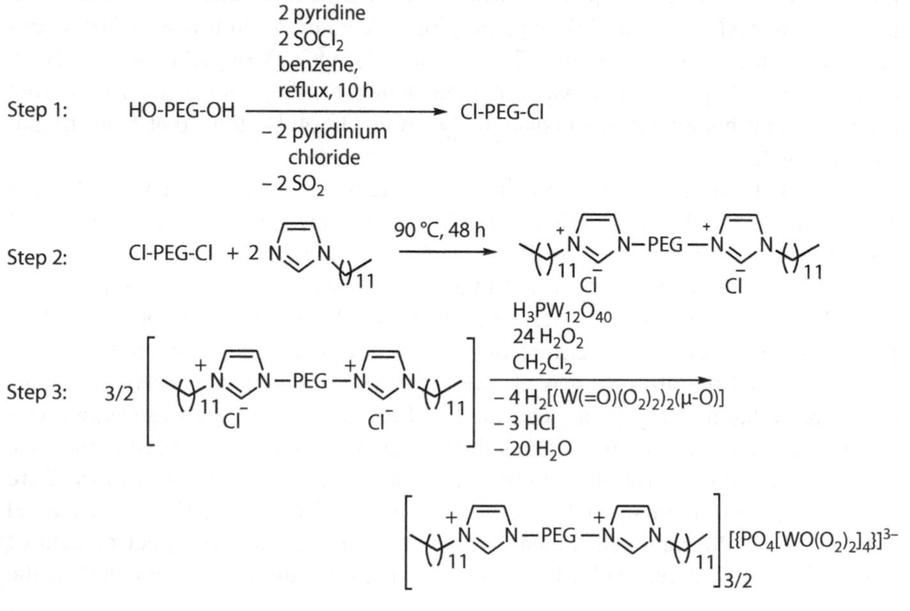

SCHEME 5.2 Preparation of POM-RTIL.

(945.6 mg, 4 mmol) were charged into a steel vessel and allowed to react at 90°C for 48 h under 1.0 MPa nitrogen with magnetic stirring. Then the product (PEG chain-functionalized N-dodecylimidazolium dichlorides) was washed with diethyl ether (1.5 mL × 4) and treated under vacuum to give the purified compound (728 mg, 0.9 mmol, 60% yield based on chlorinated PEG-300 charged). δ_H (500 MHz, CDCl$_3$): 10.68 (s, 2H, CH), 7.64 (s, 2H, CH), 7.18 (s, 2H, CH), 4.60 (t, 4H, CH$_2$), 4.21 (t, 4H, CH$_2$), 3.30–3.80 (m, PEG), 1.75 (m, 4H, CH$_2$), 1.28 (m, 36H, CH$_2$), 0.86 (t, 6H, CH$_3$). δ_C (125 MHz, CDCl$_3$): 136.7, 123.3, 121.6, 70.2–69.9, 68.8, 65.5, 49.6, 29.2, 14.9, 13.8.

Step 3: POM-RTIL catalyst. After the solution of H$_3$PW$_{12}$O$_{40}$ (1.6 g, 0.6 mmol) in 30% H$_2$O$_2$ (12 mL, 116.5 mmol) was stirred for 30 min, purified PEG chain-functionalized N-dodecylimidazolium dichlorides (0.7 g, 0.9 mmol) in dichloromethane (10 mL) were added dropwise. The resulting biphasic mixture was thoroughly stirred for another 30 min. Then the organic layer was separated and dried under anhydrous airflow at 40°C for 4 h to give an almost colorless transparent viscous ionic liquid catalyst (1.2 g, 96% yield). v_{max}/cm^{-1}: 3141, 3106, 2924, and 2854 (CH); 1736 and 1563 (CC and CN); 1465 and 1350 (CH); 1162 and 1105 (CO); 844 (OO); 523 (WOO). The bands ranged from 900 to 1200 cm^{-1} due to adsorption of P–O, and W=O were overlapped by the characteristic bands of the PEG chain. And, ^{31}P NMR (CDCl$_3$) of POM-RTIL catalyst gave mainly five signals at different chemical shifts ranging from −10 to 10 ppm, which were ascribed to the "aged" Venturello anion since it was very liable to change.[20,21]

Below are applications of POM-RTIL for the synthesis of *cis*-cyclooctene oxide (Scheme 5.3).

Method 1: Reaction under solvent-free conditions. For the reaction in the absence of solvent, POM-RTIL catalyst (50 mg, 22 μmol) was charged into a 50 mL steel vessel, followed by adding 30% H$_2$O$_2$ (340 mg, 3 mmol) and *cis*-cyclooctene (330.6 mg, 3 mmol). Then the vessel was heated to 60°C with magnetic stirring for 4 h. After the reaction, the *cis*-cyclooctene oxide was extracted with cyclohexane (1 mL × 3) for the gas chromatography (GC) analysis. The catalyst remained in the steel vessel and was treated under vacuum at 60°C for 4 h, and then reused for the next cycle (Table 5.1, entries 1–3). A 94.0% yield of *cis*-cyclooctene oxide can be obtained for the first run, and 89.1% and 88.1% for the fourth and fifth runs, respectively. When styrene (312.4 mg, 3 mmol) or cyclohexene (246.4 mg, 3 mmol) was

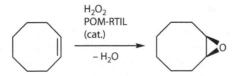

SCHEME 5.3 Preparation of *cis*-cyclooctene oxide using POM-RTIL catalyst.

TABLE 5.1
Recycling of the POM-RTIL Catalyst

Entry	Substrates	Without Solvent[a]		With Solvent[b]	
		Conversion (%)	Selectivity (%)	Conversion (%)	Selectivity (%)
1	Cyclooctene	95	99	97	99
2	Second run	91	99	97	99
3	Third run	92	99	96	99
4	Styrene	94	86[c]	93	92
5	Cyclohexene	89	23[d]	90	97

[a] Catalyst (50 mg, 22 µmol), substrate (3 mmol), 30% H_2O_2 (340 mg, 3 mmol), 60°C, 4 h.
[b] Catalyst (50 mg, 22 µmol), substrate (3 mmol), 30% H_2O_2 (113.3 mg, 1 mmol), 5 mL ethyl acetate, 60°C, 4 h. All the selectivities were found to be >99%.
[c] The main products were benzyl aldehyde and benzoic acid. The selectivity to styrene oxide was less than 2%.
[d] The 1,2-cyclohexanediol and 2-hydroxy cyclohexanone were detected as main by-products.

used as substrate in the absence of solvent, although high conversions were achieved in both cases (94% for styrene and 89% for cyclohexene), the selectivities to the epoxide were always poor due to the acidic hydrolysis of the epoxide ring by direct contact with the acid aqueous phase where acidity likely resulted from the degradation of the anion (Table 5.1, entries 4 and 5).
Method 2: Reaction in the presence of organic solvent. The reaction was carried out in ethyl acetate (5 mL) by using POM-RTIL catalyst (50 mg, 22 µmol), 30% H_2O_2 (340 mg, 3 mmol), and *cis*-cyclooctene (330.6 mg, 3 mmol). After reaction, 3 mL cyclohexane was added to the reaction mixture to facilitate the catalyst separation after the reaction. Then the solvent phase was removed, and the catalyst was washed with cyclohexane (1 mL × 3). *cis*-Cyclooctene oxide was obtained in 90.0% yield. In the case of the molar ratio of substrate (3 mmol) to 30% H_2O_2 (113 mg, 1 mmol) = 3, the yield of cyclooctene oxide was 96% for the first run after reaction at 60°C for 4 h (Table 5.1), and 90.1% and 91.1% for the fourth and fifth runs under the same conditions, respectively, which indicated that the recyclability was similar with that in the absence of ethyl acetate.

GREEN METRICS ANALYSIS

Contributed by J. Andraos

The reaction performance of the solvent-free method to prepare *cis*-cyclooctene oxide by epoxidation of *cis*-cyclooctene using hydrogen peroxide and POM-RTIL catalyst may be compared with the method employing ethyl acetate as reaction solvent, as shown in Figure 5.1. Elimination of ethyl acetate as reaction solvent reduces the process mass intensity (PMI) significantly by 61%. This result is based on a metrics

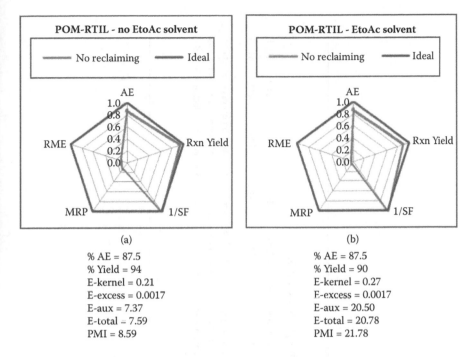

	(a)	(b)
	% AE = 87.5	% AE = 87.5
	% Yield = 94	% Yield = 90
	E-kernel = 0.21	E-kernel = 0.27
	E-excess = 0.0017	E-excess = 0.0017
	E-aux = 7.37	E-aux = 20.50
	E-total = 7.59	E-total = 20.78
	PMI = 8.59	PMI = 21.78

FIGURE 5.1 Radial pentagons showing various metrics performances for syntheses of *cis*-cyclooctene oxide from *cis*-cyclooctene using hydrogen peroxide and POM-RTIL catalyst: (a) solvent-free method and (b) method using ethyl acetate as reaction solvent.

calculation that does not include the synthesis of the POM-RTIL catalyst and assumes that it is not recovered from the waste stream. If the catalyst is recovered in both cases, the PMI values for the solvent-free and solvent-based methods are 7.45 and 20.63, respectively. Again, under this scenario, the elimination of reaction solvent reduces the PMI by 64%, indicating that the recovery of catalyst has a marginal impact on reducing overall waste compared to carrying out the reaction in the absence of solvent. Next, the waste generated from the synthesis of the catalyst was also considered in the metrics evaluation. Table 5.2 shows the metrics results for the synthesis of POM-RTIL catalyst according to the convergent plan given in Scheme 5.4.

The details of metrics calculations for the synthesis of *cis*-cyclooctene oxide that include the synthesis of POM-RTIL catalyst are summarized below. As before, it is assumed that catalysts are not reclaimed after carrying out the epoxidation reaction.

METHOD 1 (SOLVENT-FREE)

POM-RTIL catalyst loading used in *cis*-cyclooctene oxide synthesis: 0.738 mol% (0.05 g, 2.21×10^{-5} mole).

Mass of waste generated to produce 1 mole POM-RTIL = 137.35 kg
Mass of waste generated to produce 2.2×10^{-5} moles POM-RTIL = 137.35
$\times (2.2 \times 10^{-5}) = 0.00304$ kg = 3.04 g

TABLE 5.2
Summary of Metrics[a] for Synthesis of POM-RTIL Catalyst according to Scheme 5.4

Plan Type	Number of Steps	% Overall AE	% Overall Yield[b]	% Overall Kernel RME	E-Kernel	E-Excess	E-Aux	E-Total	PMI
Convergent	4	38.9	48.4	28.3	2.53	3.81	54.50	60.84	61.84

[a] Abbreviations and definitions: AE = atom economy, E-kernel = E-factor based on reaction by-products only, E-excess = E-factor based on excess reagent consumption, E-aux = E-factor based on all auxiliary materials used (reaction solvents, catalysts, workup materials, and purification materials), E-total = total E-factor (sum of above three E-factors), PMI = process mass intensity (total mass of input materials used divided by mass of target product collected as stated in Jiménez-González, C. et al, *Org. Proc. Res. Dev*, 2011, 15, 912).

[b] Calculated based on longest branch.

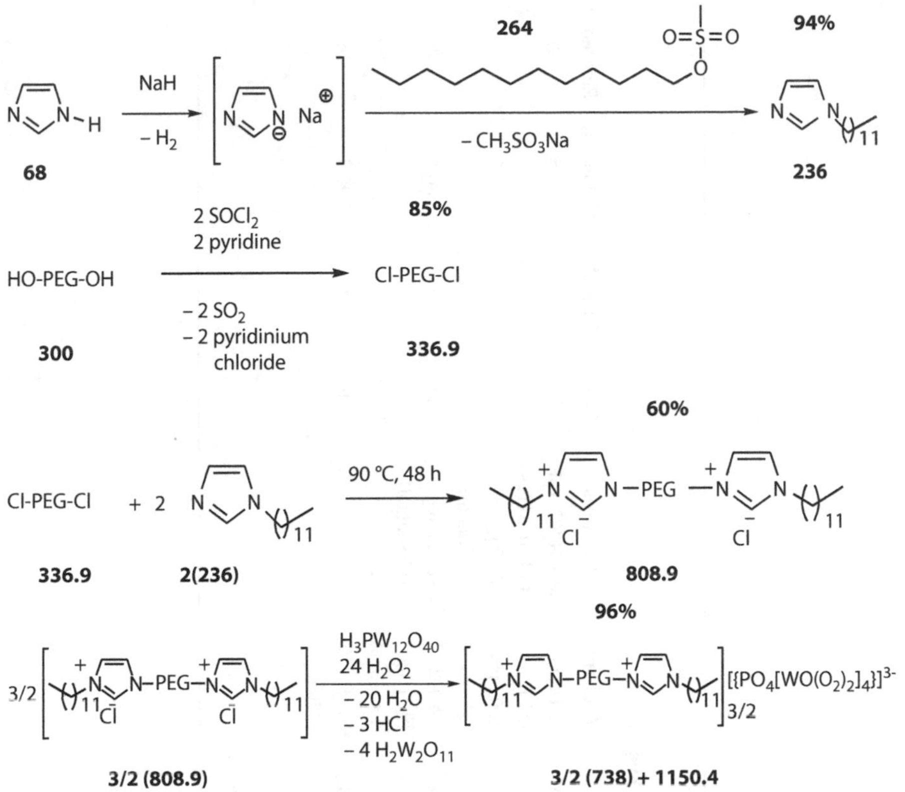

SCHEME 5.4 Convergent four-step route to POM-RTIL catalyst. Molecular weights are shown below structures.

Mass of waste generated to produce 0.36 g *cis*-cyclooctene oxide = 2.70 g
Global E-factor = (2.70 + 3.04)/0.36 = 15.94
Global PMI = 1 + 15.94 = 16.94

Method 2 (Using EtOAc as Reaction Solvent)

POM-RTIL catalyst loading used in *cis*-cyclooctene oxide synthesis: 0.72 mol% (0.05 g, 2.2×10^{-5} mole).

Mass of waste generated to produce 1 mole POM-RTIL = 137.35 kg
Mass of waste generated to produce 2.2×10^{-5} moles POM-RTIL = 137.35 $\times (2.2 \times 10^{-5}) = 0.00304$ kg = 3.04 g
Mass of waste generated to produce 0.34 g *cis*-cyclooctene oxide = 7.06 g
Global E-factor = (7.06 + 3.04)/0.34 = 29.71
Global PMI = 1 + 29.71 = 30.71

From these results it can be observed that when the synthesis of the catalyst is included in the metrics analysis for the epoxidation reaction, the global PMI for

TABLE 5.3

Comparison of Different Polyoxometalate-Based Catalytic Systems

Year	Catalytic System	Substrate	Solvent	Oxidant	Additives	T (°C)	Time (h)	Yield (%)	Reference
1983	$Na_2WO_4 \cdot 2H_2O$ (1.65g, 5 mmol), 40% w/v H_3PO_4 (2.45 mL, 10 mmol)	1-octene (22.4 g, 200 mmol)	1,2-dichloroethane (15 mL)	8% H_2O_2 (51 g, 120 mmol)	30% H_2SO_4 (pH to 1.6), MTOC (0.82 g, ca. 2 mmol)	70	0.8	82.5	6
1988	$H_3PW_{12}O_{40}$ (74.9 mg, ca. 0.026 mmol)	1-octene (728 mg, 6.5 mmol)	Chloroform (15 mL)	35% H_2O_2 (960 mg, 9.8 mmol)	CPC (28 mg, 0.078 mmol)	60	5	80.4	8
2010	POM-RTIL (50 mg, 21.6 μmol)	cis-Cyclooctene (330.6 mg, 3 mmol)	Ethyl acetate (5 mL)	30% H_2O_2 (340 mg, 3 mmol)	—	60	4	96.0	16

Note: MTOC = methyltrioctylammonium chloride, CPC = cetylpyridinium chloride.

the solvent-free method increases from 8.59 to 16.94, corresponding to a 97% increase. Similarly, the global PMI for the reaction solvent method increases from 21.78 to 30.71, which corresponds to a 41% increase. In general, inclusion of waste materials generated in the synthesis of specialized catalysts in addition to the waste generated from the reaction in question may be viewed as a fairer assessment of reaction performance.

SUMMARY

The conventional Venturello–Ishii anion-based catalytic systems and the self-separated catalyst for olefin epoxidation have been compared in detail, as shown in Table 5.3. From the table, we can see that not only was a phase-transfer agent necessary, but also the toxic organic solvent 1,2-dichloroethane was used in the conventional Venturello–Ishii system. By contrast, our method uses no additives as phase-transfer agents in the new catalytic system, and environmentally friendly ethyl acetate was used to replace the conventional toxic solvent. In addition, the POM-RTIL catalyst can be recycled at least five times without significant loss of activity. A detailed green metrics evaluation indicates that carrying out the epoxidation reaction in the absence of reaction solvent has a significant impact in reducing overall waste without sacrificing reaction yield performance. Recovery of catalyst has a minimal impact on reducing waste; however, inclusion of its synthesis results in a fairer assessment of global waste production in carrying out the epoxidation reaction. We believe that even better greener processes for the epoxidation reaction will be definitely invented in the near future.

ACKNOWLEDGMENTS

The authors are grateful for the support from the National Natural Science Foundation of China (No. 21073058), Research Fund for the Doctoral Program of Higher Education of China (20100074110014), and Fundamental Research Funds for the Central Universities, China.

REFERENCES

1. N. Mizuno, K. Yamaguchi, K. Kamata. *Coord. Chem. Rev.* 2005, 249(17–18), 1944–1956.
2. A. Proust, R. Thouvenot, P. Gouzerh. *Chem. Commun.* 2008, (16), 1837–1852.
3. D. L. Long, R. Tsunashima, L. Cronin. *Angew. Chem. Int. Ed.* 2010, 49(10), 1736–1758.
4. N. Mizuno, K. Yamaguchi, K. Kamata. *Catal. Surv. Asia* 2011, 15(2), 68–79.
5. Y. X. Qiao, Z. S. Hou. *Curr. Org. Chem.* 2009, 13(13), 1347–1365.
6. C. Venturello, E. Alneri, M. Ricci. *J. Org. Chem.* 1983, 48(21), 3831–3833.
7. C. Venturello, R. Daloisio. *J. Org. Chem.* 1988, 53(7), 1553–1557.
8. Y. Ishii, K. Yamawaki, T. Ura, H. Yamada, T. Yoshida, M. Ogawa. *J. Org. Chem.* 1988, 53(15), 3587–3593.
9. K. Kamata, S. Kuzuya, K. Uehara, S. Yamaguchi, N. Mizuno. *Inorg. Chem.* 2007, 46, 3768–3774.
10. K. Yamaguchi, C. Yoshida, S. Uchida, N. Mizuno. *J. Am. Chem. Soc.* 2005, 127(2), 530–531.

11. R. Ishimoto, K. Kamata, N. Mizuno. *Angew. Chem. Int. Ed.* 2012, 51(19), 4662–4665.
12. K. Kamata, K. Yonehara, Y. Sumida, K. Yamaguchi, S. Hikichi, N. Mizuno. *Science* 2003, 300(5621), 964–966.
13. R. Neumann, M. Gara. *J. Am. Chem. Soc.* 1995, 117, 5066–5074.
14. Z. W. Xi, N. Zhou, Y. Sun, K. L. Li. *Science* 2001, 292(5519), 1139–1141.
15. R. Alcántara, L. Canoira, P. G. Joao, J-M. Santos, I. Vázquez. *Appl. Catal. A* 2000, 203, 259–268.
16. H. Li, Z. S. Hou, Y. X. Qiao, B. Feng, Y. Hu, X. R. Wang, X. G. Zhao. *Catal. Commun.* 2010, 11(5), 470–475.
17. W. F. Hart, M. E. McGreal. *J. Org. Chem.* 1957, 22(1), 86–88.
18. V. S. Pore, N. G. Aher, M. Kumar, P. K. Shukla. *Tetrahedron* 2006, 62(48), 11178–11186.
19. Y. Qiao, Z. Hou, H. Li, Y. Hu, B. Feng, X. Wang, L. Hua, Q. Huang. *Green Chem.* 2009, 11, 1955–1960.
20. D. C. Duncan, R. C. Chambers, E. Hecht, C. L. Hill. *J. Am. Chem. Soc.* 1995, 117, 681–691.
21. I. V. Kozhevnikov, G. P. Mulder, M. C. Steverink-de Zoete, M. G. Oostwal. *J. Mol. Catal.* 1998, 134, 223–228.

6 Green Methods for the Epoxidation of Olefins Part II: Using Lacunary-Type Phosphotungstate Anion-Based Catalysts

Yunxiang Qiao and Zhenshan Hou

CONTENTS

In the area of olefin epoxidation with H$_2$O$_2$ as the oxidant, many homogeneous and heterogeneous catalytic systems have been developed.[1-5] Among them, poly-oxometalates as homogeneous catalysts have attracted much attention, particularly in the last two decades,[6-9] especially the phosphotungstate-based catalytic systems, which showed high efficiency of H$_2$O$_2$ utilization and high selectivity to the epoxides.

Lacunary derivative [PW$_{11}$O$_{39}$]$^{7-}$ was reported to be the active species for olefin epoxidation using H$_2$O$_2$ as the oxidant. Duncan et al. indicated that both H$_3$PW$_{12}$O$_{40}$ and Na$_7$PW$_{11}$O$_{39}$ were much more effective than other polyoxometalates in the epoxidation of 1-octene with a phase-transfer agent.[10] Gao et al. mentioned that the $[PW_{11}O_{39}]^{7-}$-based catalyst had reaction-controlled phase-transfer properties in cyclohexene epoxidation with 35% H_2O_2 in $CHCl_3$.[11] Recently, Wang et al. reported that only a modest activity could be obtained in the epoxidation of *cis*-cyclooctene catalyzed by $[PW_{11}O_{39}]^{7-}$ with aqueous H_2O_2 in the amphiphilic ionic liquid.[12] Weng et al. also utilized the reaction-controlled phase-transfer catalyst $[C_7H_7N(CH_3)_3]_7[PW_{11}O_{39}]$ for the oxidation of benzyl alcohol.[13]

Despite the high activity and excellent selectivity obtained in epoxidation of alkenes using the reported $[PW_{11}O_{39}]^{7-}$ catalysts, difficulty in recovery and reuse of those homogeneous metal species has restricted their wide application in industrial and laboratory synthesis. In addition, most catalytic systems used known toxic organic solvents (chloroform, 1,2-dichloroethane, benzene, acetonitrile) as reaction media.[14–17]

REACTION-CONTROLLED PHASE-TRANSFER CATALYTIC SYSTEM USING LACUNARY-TYPE CATALYST $[N-C_{16}H_{33}N(CH_3)_3]_4Na_3PW_{11}O_{39}$

Another reaction-controlled phase-transfer catalytic system composed of the lacunary-type catalyst $[n-C_{16}H_{33}N(CH_3)_3]_4Na_3PW_{11}O_{39}$, H_2O_2, olefin, and ethyl acetate was found to be highly efficient in converting olefins into the corresponding epoxides.[18] Furthermore, it was found that the lacunary-type catalyst $[n-C_{16}H_{33}N(CH_3)_3]_4Na_3PW_{11}O_{39}$ (PW_{11}) showed higher activity and better recyclability than the complete Keggin-type catalyst $[n-C_{16}H_{33}N(CH_3)_3]_3[PW_{12}O_{40}]$ (PW_{12}) under the same reaction conditions when *cis*-cyclooctene was used as the model reaction. Scheme 6.1 shows the synthesis of PW_{11} and PW_{12}, and Scheme 6.2 shows the synthesis of *cis*-cyclooctene oxide using these catalysts.

Step 1: $Na_7PW_{11}O_{39}$ was prepared according to the method reported by Brevard et al.,[19] and its structure has been confirmed by ^{31}P nuclear magnetic resonance (NMR): 22 mmol (7.25 g) of sodium tungstate dihydrate (Na_2WO_4 $2H_2O$) and 2 mmol (0.284 g) of anhydrous disodium hydrogen phosphate (Na_2HPO_4) are dissolved in 15–20 mL of water. The solution is

Step 1: $11 [Na_2WO_4 2H_2O] + Na_2HPO_4 \longrightarrow Na_7PW_{11}O_{39} + 17 NaOH + 14 H_2O$

Step 2a: $4\{[n-C_{16}H_{33}N(CH_3)_3]^+ Br^-\} + Na_7PW_{11}O_{39} \longrightarrow$

$$\{[n-C_{16}H_{33}N(CH_3)_3]_4\}^{4+} [Na_3PW_{11}O_{39}]^{4-} + 4 NaBr$$
$$(PW_{11})$$

Step 2b: $3\{[n-C_{16}H_{33}N(CH_3)_3]^+ Br^-\} + H_3PW_{12}O_{40} \longrightarrow$

$$\{[n-C_{16}H_{33}N(CH_3)_3]_3\}^{3+} [PW_{12}O_{40}]^{3-} + 3 HBr$$
$$(PW_{12})$$

SCHEME 6.1 Synthesis of PW_{11} and PW_{12} catalysts.

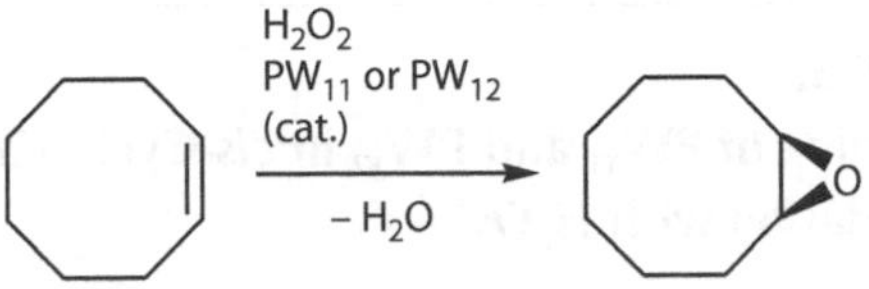

SCHEME 6.2 Synthesis of *cis*-cyclooctene oxide using PW_{11} or PW_{12} catalysts.

heated to 80–90°C and titrated exactly with concentrated nitric acid (68% in H_2O) (about 2 mL required) with vigorous stirring to pH 4.8. The volume is then reduced to half by evaporation and the heteropolyanion separated in a dense lower layer by liquid-liquid extraction with 8–10 mL of acetone. The extraction is repeated until the acetone extract shows no nitric ions (ferrous sulfate test), and then the solid sodium salt is obtained as the hydrate (15–$20H_2O$) by evaporation to dryness (in air) of the acetone extracts (4.8 g, 76% yield). δ_P (162 MHz, D_2O): -10.7 ppm.

Step 2a: Preparation of catalyst $[n\text{-}C_{16}H_{33}N(CH_3)_3]_4Na_3PW_{11}O_{39}$ (PW_{11}): $Na_7PW_{11}O_{39}$ (2.00 g, 0.7 mmol) was then dissolved in distilled water (1 mL) at 40°C with stirring, followed by addition of cetyltrimethylammonium bromide (0.77 g, 2.11 mmol) dissolved in distilled water (3 mL). A white precipitate immediately appeared and the resulting mixture was stirred for another 1 h until no more precipitate was formed. The white solid was collected, washed with distilled water (2 mL× 3) and cold ethanol (2 mL × 3), and then dried under vacuum. The white powder finally obtained is denoted PW_{11} (1.88 g, 69% yield). Analytically calculated for $C_{76}H_{168}N_4Na_3PW_{11}O_{39}$: C, 23.50; H, 4.36; N, 1.44; Na, 1.78; P, 0.80; W, 52.06. Found: C, 23.10; H, 4.40; N, 1.43; Na, 1.72; P, 0.78; W, 52.1. v_{max}/cm^{-1}: 1081 and 1047 (P–O), 952 (W–O_d), 894 and 857 (W–O_b–W), 803 and 730 (W–Oc–W).

Step 2b: The quaternary ammonium salt $[n\text{-}C_{16}H_{33}N(CH_3)_3]_3PW_{12}O_{40}$ (PW_{12}) was prepared by the same procedure as reported previously[20]: To a solution of cetyltrimethylammonium bromide (1.87 g, 5.2 mmol) in 20 mL of water was added dropwise $H_3PW_{12}O_{40}$ (4.9 g, ca. 1.7 mmol) in 10 mL of water with stirring at ambient temperature to form a white precipitate immediately. After being stirred continuously for 3–4 h, the resulting mixture was filtered, then washed with distilled water (10 mL × 5), and dried in vacuum to give PW_{12} as white powder (4.9 g, 77% yield). Analytically calculated for $C_{57}H_{126}N_3PW_{12}O_{40}$: C, 18.35; H, 3.40; N, 1.13; P, 0.83; W, 59.13. Found: C, 18.95; H, 3.70; N, 1.53; P, 0.86; W, 61.00. The structure of catalyst was also confirmed by Fourier transform infrared (FT-IR) and ^{31}P NMR.[19]

PREPARATION OF CIS-CYCLOOCTENE OXIDE

The reaction was carried out in a 25 mL Schlenk flask equipped with a reflux condenser and thermometer. To 2 mL of ethyl acetate was added PW_{11} (56 mg, 0.0144 mmol)

TABLE 6.1
Recycling of PW₁₁ and PW₁₂ in *cis*-Cyclooctene Epoxidation with H₂O₂

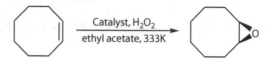

Run	Conversion (%)	
	PW$_{11}$	PW$_{12}$
1	100	100
2	100	100
3	100	100

or PW$_{12}$ (54 mg, 0.0144 mmol), *cis*-cyclooctene (0.2 mL, 1.54 mmol), and 30% H$_2$O$_2$ (0.088 g, 0.77 mmol). The reaction mixture was stirred at 60°C for 1 h. The reaction mixture was cooled to 0°C, and then the catalyst was precipitated and separated. The catalyst was used for the next catalytic recycle after washing with cold ethyl acetate (1 mL × 2) and drying at 50°C for 2 h under reduced pressure. The gas chromatography (GC) yield of *cis*-cyclooctene oxide was found to be >99% with >99% selectivity.

The catalyst PW$_{11}$ showed excellent activity for epoxidation of *cis*-cyclooctene using 30% H$_2$O$_2$ as oxidant (Table 6.1). Like other reaction-controlled phase-transfer catalysts, PW$_{11}$ can be transferred to the organic phase by reaction with H$_2$O$_2$ and then precipitated after H$_2$O$_2$ is consumed. The catalyst PW$_{11}$ can be employed at least 10 times without decreasing the catalytic activity in the epoxidation of *cis*-cyclooctene.

REACTION CONDITIONS

0.0144 mmol catalyst (56 mg for PW$_{11}$, 54 mg for PW$_{12}$), *cis*-cyclooctene (0.2 mL, 1.54 mmol), H$_2$O$_2$ (0.088 g, 0.77 mmol), 2 mL ethyl acetate, 60°C, 1 h. The selectivity to *cis*-cyclooctene oxide was always over 99%.

In addition, the lacunary-type phosphotungstate anion-based catalyst PW$_{11}$ gave a better catalytic performance than the complete Keggin-type anion-based catalyst PW$_{12}$ in ethyl acetate from the point of reaction dynamics. As shown in Figure 6.1, it was very clear that the reaction rate of catalyst PW$_{11}$ was much higher than that of catalyst PW$_{12}$ in ethyl acetate media. ^{31}P NMR spectroscopy and solubility measurements for the two catalysts revealed that the [PW$_{11}$O$_{39}$]$^{7-}$ anion had a much faster degradation rate than the [PW$_{12}$O$_{40}$]$^{3-}$ anion in an excess of H$_2$O$_2$, which resulted in the formation of more catalytically active species. Therefore, catalyst PW$_{12}$ showed 98% conversion for the fourth run for *cis*-cyclooctene epoxidation, and it dropped to 78% for the fifth run.

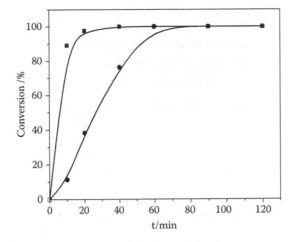

FIGURE 6.1 Time profile of epoxidation of *cis*-cyclooctene catalyzed by PW_{11} (■) and PW_{12} (●). Reaction conditions: 0.0144 mmol catalyst, 1.54 mmol *cis*-cyclooctene, 0.77 mmol H_2O_2, 2 ml ethyl acetate, 60°C. The selectivity to *cis*-cyclooctene oxide was always over 99%.

TI-SUBSTITUTED PHOSPHOTUNGSTATE ([DMIM]₅PTIW₁₁O₄₀) AS A HETEROGENEOUS CATALYST FOR OLEFIN EPOXIDATION

Additionally, a Ti-substituted phosphotungstate ([DMIM]$_5$PTiW$_{11}$O$_{40}$) (DMIM = 1-dodecyl-3-methyl-imidazolium) was also found to be a heterogeneous catalyst for olefin epoxidation with aqueous hydrogen peroxide.[21] The heterogeneous nature of catalysis in ethyl acetate media was also confirmed by a hot catalyst filtration test. The catalytic system can convert different kinds of olefins to the corresponding epoxides. In the epoxidation reaction, the Ti-based peroxo structure was believed to be the active species, which was quite stable and maintained the peroxo structure even after the reaction, thus enhancing the rate of the catalytic reaction in the following run due to the disappearance of the induction period. The heterogeneous catalyst was easily separated and recycled eight times without decreasing remarkably the catalytic activity (Table 6.2).

Ti-substituted phosphotungstate [DMIM]$_5$PTiW$_{11}$O$_{40}$ is prepared as shown in Scheme 6.3, and its application for the synthesis of *cis*-cyclooctene oxide is shown in Scheme 6.4.

Step 1: The 1-dodecyl-3-methylimidazolium bromide ([DMIM]Br) was synthesized according to the conventional method.[22] 1-Methyl imidazole (1.07 g, 13 mmol) was allowed to react with dodecyl bromide (3.22 g, 13 mmol) at 90°C for 48 h in an autoclave under a 0.7 MPa N_2 atmosphere. After the reaction, the products were washed with diethyl ether (5 mL × 4) and then dried at 50°C under vacuum for 1 h (4.12 g, 96% yield). δ_H (400 MHz, D_2O): 0.758 (3H, t), 1.164 (18H, m), 1.818 (2H, m), 3.868 (3H, s), 4.196 (2H, t), 7.489 (1H, s), 7.498 (1H, s), 8.919 (1H, s).

TABLE 6.2

Recyclability of the Catalyst [DMIM]$_5$PTiW$_{11}$O$_{40}$ for *cis*-Cyclooctene Epoxidation

Run	Conversion (%)	Selectivity (%)
1	>99%	>99%
2	>99%	>99%
3	>99%	>99%

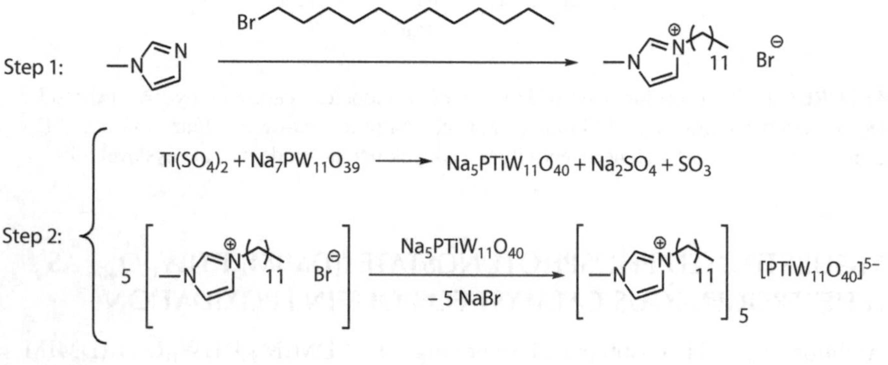

Step 1:

Step 2:

$$Ti(SO_4)_2 + Na_7PW_{11}O_{39} \longrightarrow Na_5PTiW_{11}O_{40} + Na_2SO_4 + SO_3$$

SCHEME 6.3 Preparation of [DMIM]$_5$PTiW$_{11}$O$_{40}$ catalyst.

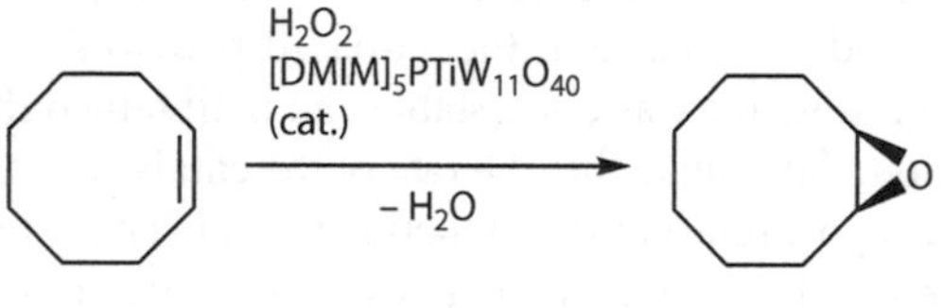

SCHEME 6.4 Synthesis of *cis*-cyclooctene oxide using [DMIM]$_5$PTiW$_{11}$O$_{40}$ catalyst.

Step 2: The preparation of [DMIM]$_5$PTiW$_{11}$O$_{40}$ was similar to the procedure reported by Kholdeeva et al.[23] Briefly, Ti(SO$_4$)$_2$ (0.169 g, 0.7 mmol; 1 mol/L solution in 2 mol/L H$_2$SO$_4$) was added to an aqueous solution (2 mL) of Na$_7$PW$_{11}$O$_{39}$ (2 g, 0.7 mmol). After the pH of the resulting solution was adjusted to 5.5 by adding small aliquots of NaHCO$_3$, the mixed solution was stirred at room temperature for another 12 h. The catalyst was then precipitated by adding [DMIM]Br (1.15 g, 3.5 mmol). The white solids were collected, washed with distilled water (3 mL × 3), and then dried under vacuum (1.75 g, 62% yield). The white powder was obtained as [DMIM]$_5$PTiW$_{11}$O$_{40}$. Analytically calculated for C$_{80}$H$_{155}$N$_{10}$PTiW$_{11}$O$_{40}$: C, 24.03; H, 3.91; N, 3.50;

Ti, 1.20; W, 50.58%. Found: C, 23.44; H, 3.38; N, 3.40; Ti, 1.38; W, 55.3%. δ_P (161.9 MHz, CD$_3$CN): –13.97ppm. ν_{max}/cm^{-1}: 1089, 1064, 961, 889, 804, 704, and 592.

PREPARATION OF *CIS*-CYCLOOCTENE OXIDE

The reaction was carried out in a 25 mL Schlenk flask equipped with a reflux condenser and a thermometer. To 2 mL ethyl acetate was added catalyst [DMIM]$_5$PTiW$_{11}$O$_{40}$ (60 mg, 0.015 mmol), *cis*-cyclooctene (0.17 g, 1.54 mmol), and 30% H$_2$O$_2$ (0.085 g, 0.75 mmol). The reaction mixture was stirred at 60°C for 4 h. Then the reaction mixture was cooled down to 0°C, and the catalyst was precipitated and separated. After being washed with cold ethyl acetate (1 mL × 2) and dried at 50°C for 2 h under reduced pressure, the catalyst was then used for the next catalytic recycle (Table 6.2).

REACTION CONDITIONS

[DMIM]$_5$PTiW$_{11}$O$_{40}$ (60 mg, 0.015 mmol), *cis*-cyclooctene (170 mg, 1.54 mmol), H$_2$O$_2$ (85 mg, 0.75 mmol), ethyl acetate (2 mL), 60°C, 4 h. Conversion of olefins = the amount of olefins converted (moles)/initial H$_2$O$_2$ (moles) ×100%.

GREEN METRICS ANALYSIS

Contributed by J. Andraos

The reaction performances of the three catalysts PW$_{11}$, PW$_{12}$, and [DMIM]$_5$[PTiW$_{11}$O$_{40}$] for the preparation of *cis*-cyclooctene oxide by epoxidation of *cis*-cyclooctene using hydrogen peroxide are summarized in Figure 6.2. All three reactions have similar

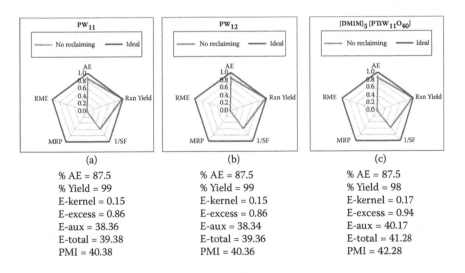

(a)	(b)	(c)
% AE = 87.5	% AE = 87.5	% AE = 87.5
% Yield = 99	% Yield = 99	% Yield = 98
E-kernel = 0.15	E-kernel = 0.15	E-kernel = 0.17
E-excess = 0.86	E-excess = 0.86	E-excess = 0.94
E-aux = 38.36	E-aux = 38.34	E-aux = 40.17
E-total = 39.38	E-total = 39.36	E-total = 41.28
PMI = 40.38	PMI = 40.36	PMI = 42.28

FIGURE 6.2 Radial pentagons showing various metrics performances for syntheses of *cis*-cyclooctene oxide from *cis*-cyclooctene using (a) PW$_{11}$ catalyst, (b) PW$_{12}$ catalyst, and (c) [DMIM]$_5$[PTiW$_{11}$O$_{40}$] catalyst.

process mass intensity (PMI) values and yield performances and E-factor profiles. The main contributor to overall waste is the ethyl acetate solvent that is used in the reaction and the workup. As was done for Part 1 (Chapter 5), these calculations are based on the assumption that catalysts are not recovered from the waste stream. If all catalysts are recovered, the respective PMI values are 38.80, 38.80, and 40.63, representing a reduction of 3.9% in each case. Next, the waste generated from the syntheses of both catalysts was also considered in the metrics evaluation. Table 6.3 summarizes the metrics results for their syntheses according to the plans given in Schemes 6.5 to 6.7.

The details of metrics calculations for the syntheses of *cis*-cyclooctene oxide that include the synthesis of catalysts are summarized below. As before, it is assumed that catalysts are not reclaimed after carrying out the epoxidation reaction.

REACTION A

PW_{11} catalyst loading used in *cis*-cyclooctene oxide synthesis: 1.86 mol% (0.056 g, 1.44×10^{-5} mole).

Mass of waste generated to produce 1 mole PW_{11} = 68.79 kg
Mass of waste generated to produce 1.44×10^{-5} moles PW_{11} = $68.79 \times (1.44 \times 10^{-5})$
= 0.000991 kg = 0.991 g
Mass of waste generated to produce 0.097 g *cis*-cyclooctene oxide = 3.82 g
Global E-factor = (3.82 + 0.991)/0.097 = 49.60
Global PMI = 1 + 49.60 = 50.60

REACTION B

PW_{12} catalyst loading used in *cis*-cyclooctene oxide synthesis: 1.86 mol% (0.054 g, 1.45×10^{-5} mole).

Mass of waste generated to produce 1 mole PW_{12} = 62.31 kg
Mass of waste generated to produce 1.45×10^{-5} moles PW_{12} = $62.31 \times (1.45 \times 10^{-5})$
= 0.000904 kg = 0.904 g
Mass of waste generated to produce 0.097 g *cis*-cyclooctene oxide = 3.82 g
Global E-factor = (3.82 + 0.904)/0.097 = 48.70
Global PMI = 1 + 48.70 = 49.70

REACTION C

$[DMIM]_5PTiW_{11}O_{40}$ catalyst loading used in *cis*-cyclooctene oxide synthesis: 2 mol% (0.06 g, 1.5×10^{-5} mole).

Mass of waste generated to produce 1 mole $[DMIM]_5PTiW_{11}O_{40}$ = 75.74 kg
Mass of waste generated to produce 1.5×10^{-5} moles $[DMIM]_5PTiW_{11}O_{40}$
= $75.74 \times (1.5 \times 10^{-5})$ = 0.00114 kg = 1.14 g

TABLE 6.3
Summary of Metrics[a] for Syntheses of Catalysts according to Schemes 6.5 to 6.7

Catalyst	Number of Steps	% Overall AE	% Overall Yield[b]	% Overall Kernel RME	E-Kernel	E-Excess	E-Aux	E-Total	PMI
PW_{11}	2	74.0	53.8	42.4	1.36	9.31	7.05	17.72	18.72
PW_{12}	1	84.8	77.2	72.4	0.38	0.0026	16.33	16.71	17.71
$[DMIM]_5$ $PTiW_{11}O_{40}$	3	70.3	48.4	36.3	1.75	5.16	12.04	18.95	19.95

[a] Abbreviations and definitions: AE = atom economy, E-kernel = E-factor based on reaction by-products only, E-excess = E-factor based on excess reagent consumption, E-aux = E-factor based on all auxiliary materials used (reaction solvents, catalysts, workup materials, and purification materials), E-total = total E-factor (sum of above three E-factors), PMI = process mass intensity (total mass of input materials used divided by mass of target product collected as stated in Jiménez-González, C. et al., *Org. Proc. Res. Dev*, 2011, 15, 912).

[b] Calculated based on longest branch.

78%

$$11 \, [Na_2WO_4 \, 2 \, H_2O] + Na_2HPO_4 + H_2O \longrightarrow Na_7PW_{11}O_{39} \, 15 \, H_2O + 17 \, NaOH$$

11 (329.85)   142   18   2838.35 + 15*18   17(40)

69%

$$4 \{[n\text{-}C_{16}H_{33}N(CH_3)_3]^+ \, Br^-\} + Na_7PW_{11}O_{39} \, 15 \, H_2O \xrightarrow{\;-15 \, H_2O\;} \{[n\text{-}C_{16}H_{33}N(CH_3)_3]_4\}^{4+} \, [Na_3PW_{11}O_{39}]^{4-} + 4 \, NaBr$$

(PW₁₁)

4 (363.9)   2838.35 + 15*18   3882.35   4 (102.9)

SCHEME 6.5   Linear two-step route to PW$_{11}$ catalyst. Molecular weights are shown below structures.

77%

$$3 \{[n\text{-}C_{16}H_{33}N(CH_3)_3]^+Br^-\} + H_3PW_{12}O_{40} \longrightarrow \{[n\text{-}C_{16}H_{33}N(CH_3)_3]_3\}^{3+}[PW_{12}O_{40}]^{3-} + 3 \, HBr$$

(PW₁₂)

3 (363.9)   2880.2   3729.2   3 (80.9)

SCHEME 6.6   Linear one-step route to PW$_{12}$ catalyst. Molecular weights are shown below structures.

248.9   96%

82   330.9

78 %

$$11 \, [Na_2WO_4 \, 2 \, H_2O] + Na_2HPO_4 + H_2O \longrightarrow Na_7PW_{11}O_{39} \, 15 \, H_2O + 17 \, NaOH$$

11 (329.85)   142   18   2838.35 + 15*18   17 (40)

$$Ti(SO_4)_2 + Na_7PW_{11}O_{39} \, 15 \, H_2O \longrightarrow [Na_5PTiW_{11}O_{40}] + Na_2SO_4 + SO_3 + 15 \, H_2O$$

239.9   2838.35 + 15*18   2856.25   142   80   15 (18)

62%

$$5 \left[-\text{N} \overset{\oplus}{\underset{11}{\diagup}} \text{N} \diagdown\diagup \quad Br^{\ominus} \right] \xrightarrow[\;-5 \, NaBr\;]{[Na_5PTiW_{11}O_{40}]} \left[-\text{N} \overset{\oplus}{\underset{11}{\diagup}} \text{N} \diagdown\diagup \right]_5 [PTiW_{11}O_{40}]^{5-}$$

5 (330.9)   3996.25

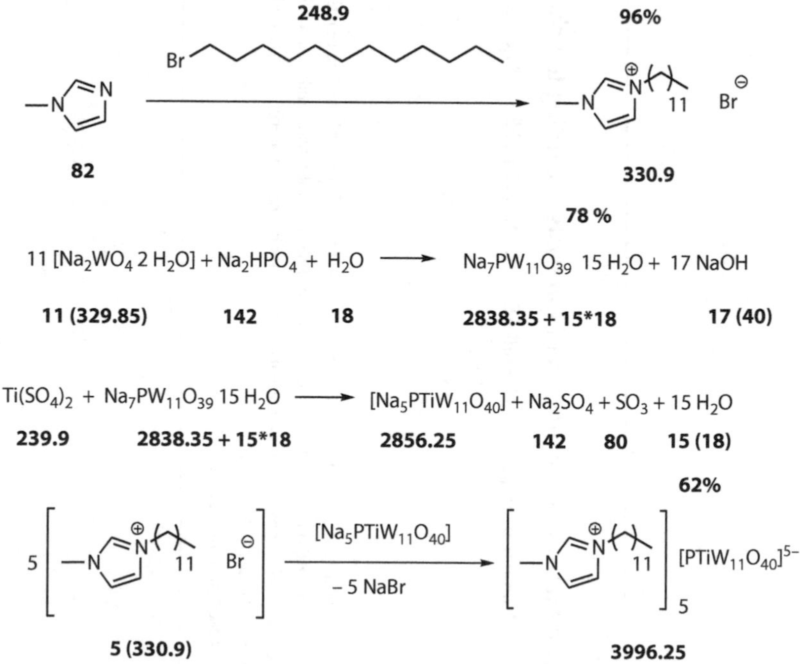

SCHEME 6.7   Convergent three-step route to [DMIM]$_5$[PTiW$_{11}$O$_{40}$] catalyst. Molecular weights are shown below structures.

Mass of waste generated to produce 0.093 g *cis*-cyclooctene oxide = 3.83 g
Global E-factor = (1.14 + 3.83)/0.093 = 53.44
Global PMI = 1 + 53.44 = 54.44

From these results we find that when the synthesis of PW_{11} is included in the metrics analysis for the epoxidation of *cis*-cyclooctene, the global PMI increases from 40.38 to 50.60, corresponding to a 25% increase. Similarly for the reactions involving PW_{12} and $[DMIM]_5[PTiW_{11}O_{40}]$ catalysts, the global PMI increases from 40.36 to 49.70, and from 42.28 to 54.44, corresponding to increases of 23% and 29%, respectively. The PMI increment due to inclusion of PW_{11} and PW_{12} syntheses is lower than that of $[DMIM]_5PTiW_{11}O_{40}$ mainly because they are made in fewer steps, thus decreasing overall waste production. This is consistent with the result shown in Table 6.3.

For all epoxidations of *cis*-cyclooctene investigated in Parts 1 and 2 we then examined how they rank in metrics performance with previously published procedures. Due to the wide variety of catalysts employed, in all calculations the syntheses of catalysts were not included so that the performance of the core epoxidation reaction could be compared head-to-head. Tables 6.4 to 6.8 summarize the key metrics parameters, including turnover numbers (TONs) and whether catalysts were recovered or not, for epoxidations involving hydrogen peroxide, *tert*-butyl hydroperoxide, peracids, molecular oxygen, and miscellaneous oxidants. Table 6.9 summarizes the best-performing epoxidation reactions using each kind of oxidant. From these exhaustive results it appears that the present methodologies described in Parts 1 (preceding chapter) and 2 are very competitive with what has been reported in the literature. With respect to hydrogen peroxide as oxidant, our procedure using the POM-RTIL catalyst ranks second overall in PMI performance, behind the industrial Badische Anilin-und-Soda-Fabrik (BASF) procedure using formic acid.[24] However, there is further room for improvement for lacunary-type phosphotungstate anion-based catalysts.

SUMMARY

The phase-transfer catalyst $[n\text{-}C_{16}H_{33}N(CH_3)_3]_4Na_3PW_{11}O_{39}$ (PW_{11}) was developed and showed higher activity for epoxidation than the complete Keggin-type anion-based catalyst ($[n\text{-}C_{16}H_{33}N(CH_3)_3]_3PW_{12}O_{40}$). It could be recycled at least 10 times without decreasing activity, and solved the difficulty in recovery and reuse of those homogeneous metal species to some extent. In addition, Ti-substituted phosphotungstate $[DMIM]_5PTiW_{11}O_{40}$ was also found to be a heterogeneous catalyst for olefin epoxidation with aqueous hydrogen peroxide. It can be recycled eight times without decreasing remarkably the catalytic activity. Environment-friendly ethyl acetate was used as a reaction media instead of conventional toxic organic solvents like toluene, acetonitrile, benzene, etc. However, its elimination as reaction solvent can significantly reduce overall waste, as shown by the metrics results of reaction performances between PW_{11}, PW_{12}, and $[DMIM]_5PTiW_{11}O_{40}$ catalysts. Among other reported procedures for the epoxidation of *cis*-cyclooctene using hydrogen peroxide as oxidant, the methodologies presented in Parts 1 and 2 of this work rank competitively with respect to reaction greenness based on material efficiency.

TABLE 6.4
Green Metrics Summary of *cis*-Cyclooctene to *cis*-Cyclooctene Oxide Reactions Using Hydrogen Peroxide as Oxidant[a]

Plan	Catalyst[b]	Recycling Catalyst?	TON	% AE	% Yield	E-Kernel	E-Excess	E-Aux	E-Total	PMI
BASF[q]	Formic acid	No	2	87.5	76	0.50	0.10	4.52	5.13	6.13
Hou (this work, Part 1)	**POM-RTIL (No EtOAc reaction solvent)**	**Yes**	**128**	**87.5**	**94**	**0.21**	**0.0017**	**7.37**	**7.59**	**8.59**
Inoue[g]	MoO$_2$(acac)$_2$ (nBu$_3$Sn)$_2$O	No	157	87.5	98	0.17	0.42	7.48	8.07	9.07
Feringa[dd]	[Mn$_2$O$_3$(L10)$_2$][PF$_6$]$_2$ Sodium oxalate GMHA	No	820	87.5	82	0.39	0.099	8.28	8.77	9.77
Louloudi[h]	MnCl$_2$(L1)	No	778	87.5	78	0.47	0.35	9.27	10.09	11.09
Louloudi[v]	Mn(OAc)$_2$(L7)	Yes	710	87.5	71	0.61	0.38	10.16	11.15	12.15
Terent'ev[y]	BF$_3$ OEt$_2$	No	0.7	87.5	73	0.57	0.012	12.91	13.49	14.49
Pescarmona[l]	Ga$_2$O$_3$	Yes	64	87.5	64	0.78	0.42	14.38	15.59	16.59
Bogdal[aa]	Na$_2$WO$_4$ 2H$_2$O [CH$_3$(C$_8$H$_{17}$)$_3$N]Cl	No	37	87.5	97	0.18	1.11	16.92	18.22	19.22
Hou (this work, Part 1)	**POM-RTIL (EtOAc as reaction solvent)**	**Yes**	**125**	**87.5**	**90**	**0.27**	**0.0017**	**20.50**	**20.78**	**21.78**
Hou[w]	[C$_{12}$mim]$_5$PTiW$_{11}$O$_{40}$	Yes	49	87.5	98	0.17	0.94	20.73	21.84	22.84
Que[s]	[Fe(II)(L6)(CH$_3$CN)$_2$](OTf)$_2$	No	8.5	87.5	85	0.34	19.51	7.73	27.59	28.59
Bach[c]	KHCO$_3$	No	3	87.5	60	0.90	0.021	29.05	29.97	30.97
Hou (this work, Part 2)	**PW$_{12}$**	**Yes**	**53**	**87.5**	**99**	**0.15**	**0.86**	**38.34**	**39.36**	**40.36**

Hou (this work, Part 2)	**PW$_{11}$**	**Yes**	**53**	**87.5**	**99**	**0.15**	**0.86**	**38.36**	**39.38**	**40.38**
Hou (this work, Part 2)	**[DMIM]$_5$PTiW$_{11}$O$_{40}$**	**Yes**	**49**	**87.5**	**98**	**0.17**	**0.94**	**40.17**	**41.28**	**42.28**
Saladino[bb]	ReO$_3$(CH$_3$)(L9)	Yes	98	87.5	98	0.17	0.14	43.63	43.93	44.93
Feringa[r]	[(N3Py-Bn) Fe(CH$_3$CN)$_2$] (ClO4)$_2$	No	16	87.5	32	2.57	51.84	2.32	56.72	57.72
Tamami[f]	[Co]	Yes	38	87.5	75	0.52	1.44	55.50	57.46	58.46
Stark[cc]	Na$_2$HPO$_4$, 1,1,3,3-tetrachloro-propan-2-one	Yes	5	87.5	67	3.03	0.43	62.48	65.94	66.94
Feringa[j]	Mn$_2$O(OAc)$_2$(TPTN)	No	262	87.5	26	3.36	7.21	58.41	68.98	69.98
Notestein[e]	Tantalum calixarene catalyst	No	18	87.5	16	6.07	2.74	69.30	78.10	79.10
Sen[k]	Calcined vanado silicate catalyst	No	3019	87.5	10	10.98	0.22	79.40	90.60	91.60
Pescarmona[t]	Ga-MCM41	Yes	0.5	87.5	20	4.63	11.95	102.22	118.80	119.80
Louloudi[i]	FeCl$_3$(L2)	Yes	7	87.5	37	2.09	115.54	2.43	120.06	121.06
Beller[o]	FeCl$_3$ 6H$_2$O (L5)	No	13	87.5	65	0.76	0.97	196.67	198.40	199.40
Tangestaninejad[u]	[n-Bu$_4$N]$_5$ [PZnMo$_2$W$_9$O$_{39}$]	Yes	1830	87.5	85	0.34	2.85	196.60	199.79	200.79

Continued

TABLE 6.4 (Continued)
Green Metrics Summary of cis-Cyclooctene Oxide Reactions Using Hydrogen Peroxide as Oxidant[a]

Plan	Catalyst[b]	Recycling Catalyst?	TON	% AE	% Yield	E-Kernel	E-Excess	E-Aux	E-Total	PMI
Feringa[x]	Mn(ClO$_4$)$_2$, 6H$_2$O Pyridine-2-carboxylic acid	No	611	87.5	57	1.01	0.19	274.20	275.39	276.39
Que[d]	Fe(II)tpp	No	172	87.5	17	5.64	312.20	250.36	568.21	569.21
Costas[n]	[Fe(II)(L4)](OTf)$_2$	No	49	87.5	99	1.47	57.72	149.51	208.70	209.70
Costas–Que[o]	[Fe(II)(L4)](OTf)$_2$	No	7	87.5	65	0.76	132.93	5340.57	5474.26	5475.26
Gebbink[m]	[Fe(II)(L3)$_2$](OTf)$_2$	No	4	87.5	0.4	299.75	1349.21	5083.05	6732.00	6733.00
Rutledge[z]	Fe(OAc)$_2$, Na$_2$(L8)	No	0.02	87.5	0.2	614.13	46519	944988	992121	992122

Note: Metrics do not include preparation of catalysts.

[a] Abbreviations and definitions: AE = atom economy, E-kernel = E-factor based on reaction by-products only, E-excess = E-factor based on excess reagent consumption, E-aux = E-factor based on all auxiliary materials used (reaction solvents, catalysts, workup materials, and purification materials), E-total = total E-factor (sum of above three E-factors), PMI = process mass intensity (total mass of input materials used divided by mass of target product collected as stated in Jiménez-González, C. et al., *Org. Proc. Res. Dev.*, 2011, 15, 912), TON = turnover number = moles of product/moles of catalyst.

[b] Catalyst definitions: C$_{12}$mim = 3-dodecyl-1-methyl-3H-imidazol-1-ium, [Co] = see reference 6, GMHA = glyoxylic acid methyl ester methyl hemiacetal, L1 = (1H-pyrrol-3-ylmethylene)-[2-(3-{2-[((1H-pyrrol-3-ylmethylene)-amino]-ethyl}-2,3,4,5-tetrahydro-3'H-[2,4']biimidazolyl-1-yl)-ethyl]-amine, L2 = 3-{2-[2-(3-Hydroxy-1,3-diphenyl-propylideneamino)-ethylamino]-ethylimino}-1,3-diphenyl-propan-1-ol, L3 = 3,3-bis-(1-methyl-1H-imidazol-2-yl)-propionic acid propyl ester, L4 = 1,4-dimethyl-7-pyridin-2-ylmethyl-[1,4,7]triazonane, L5 = 1-(2,6-diisopropyl-phenyl)-1H-imidazole, N3Py-Bn = N-benzyl-N-[di(2- pyridinyl)methyl]-N-(2-pyridinylmethyl)amine, L6 = N,N'-dimethyl-N,N'-bis-pyridin-2-ylmethyl-ethane-1,2-diamine, L7 = 4-(2-{3-[2-(1-methyl-3-oxo-butylideneamino)-ethyl]-2-[4-(3-trihydroxysilanyl-propoxy)-phenyl]-imidazolidin-1-yl}-ethylimino)-pentan-2-one,L8=3-{6-[(ethyl-phenyl-amino)-methyl]-pyridin-2-yl}-2-hydroxy-2-phenyl-propionic acid, L9 = pyridin-2-yl-methylamine, L10 = 1,4,7-trimethyl-1,4,7-triazacyclononane, POM-RTIL = polyoxometalate room temperature ionic liquid, PW$_{11}$ = [n-C$_{16}$H$_{33}$N(CH$_3$)$_3$]$_4$Na$_3$PW$_{11}$O$_{39}$, TPTN = N,N,N',N'-tetrakis(2-pyridylmethyl)propane-1,3-diamine, tpp = *meso*-tetrakis(pentafluorophenyl) porphyrin.

[c] Bach, R. D., Knight, J. W., *Org. Synth.*, 1981, 60, 63.

[d] Chen, K., Que, L., Jr., *Angew. Chem. Int. Ed.*, 1999, 38, 2227.

[e] Morlanes, N., Notestein, J. M., *Appl. Catal., A* 2010, 387, 45.

[f] Tamami, B., Ghasemi, S., *Appl. Catal., A* 2011, 393, 242.

[g] Kamiyama, T., Inoue, M., Kashiwagi, H., Enomoto, S., *Bull. Chem. Soc. Jpn.*, 1990, 63, 1559.

[h] Stamatis, A., Vartzouma, C., Louloudi, M., *Catal. Commun.*, 2011, 12, 475.

[i] Bilis, G., Christoforidis, K. C., Deligiannakis, Y., Louloudi, M., *Catal. Today*, 2010, 157, 101.

[j] Brinksma, J., Hage, R., Kerschnerc, J., Feringa, B. L., *Chem. Commun.*, 2000, 537.

[k] Sen, T., Whittle, J., Howard, M., *Chem. Commun.*, 2012, 4232.

[l] Pescarmona, P. P., Janssen, K. P. F., Jacobs, P. A., *Chem. Eur. J.*, 2007, 13, 6562.

[m] Bruijnincx, P. C. A., Buurmans, I. L. C., Gosiewska, S., Moelands, M. A. H., Lutz, M., Spek, A. L., Koten, G., Gebbink, R. J. M. K., *Chem. Eur. J.*, 2008, 14, 1228.

[n] Company, A., Gomez, L., Fontrodona, X., Ribas, X., Costas, M., *Chem. Eur. J.*, 2008, 14, 5727.

[o] Company, A., Feng, Y., Guell, M., Ribas, X., Luis, J. M., Que, L., Jr., Costas, M., *Chem. Eur. J.*, 2009, 15, 3359.

[p] Schroder, K., Enthaler, S., Bitterlich, B., Schulz, T., Spannenberg, A., Tse, M. K., Junge, K., Beller, M., *Chem. Eur. J.*, 2009, 15, 5471.

[q] Pachaly, H., Schlichting, O., DE 962073 (BASF, 1955).

[r] Klopstra, M., Roelfes, G., Hage, R., Kellogg, R. M., Feringa, B. L., *Eur. Inorg. Chem.*, 2004, 846.

[s] Mas-Balleste, R., Que, L., Jr., *J. Am. Chem. Soc.*, 2007, 129, 15964.

[t] Pescarmona, P. P., van Noyen, J., Jacobs, P. A., *J. Catal.*, 2007, 251, 307.

[u] Saedi, Z., Tangestaninejad, S., Moghadam, M., Mirkhani, V., Mohammadpoor-Baltork, I., *J. Coordin. Chem.*, 2012, 65, 463.

[v] Stamatis, A., Giasafaki, D., Christoforidis, K. C., Deligiannakis, Y., Louloudi, M., *J. Mol. Catal. A* 2010, 319, 58.

[w] Hua, L., Qiao, Y., Yu, Y., Zhu, W., Cao, T., Shi, Y., Li, H., Feng, B., Hou, Z., *New J. Chem.*, 2011, 35, 1836.

[x] Saisah, P., Pijper, D., van Summeren, R. P., Hoen, R., Smit, C., de Boer, J. W., Hage, R., Alsters, P. L., Feringa, B. L., Browne, W. R., *Org. Biomol. Chem.*, 2010, 8, 4444.

[y] Terent'ev, A. O., Boyarinova, K. A., Nikishin, G. I., *Russ. J. Gen. Chem.*, 2008, 78, 592.

[z] Barry, S. M., Rutledge, P. J., *Synlett*, 2008, 2172.

[aa] Bogdal, D., Lukasiewicz, M., Pielichowski, J., Bednarz, S., *Synth. Commun.*, 2005, 35, 2973.

[bb] Saladino, R., Andreoni, A., Neria, V., Crestini, C., *Tetrahedron*, 2005, 61, 1069.

[cc] Stark, C. J., *Tetrahedron Lett.*, 1981, 22, 2089.

[dd] Brinksma, J., Schmieder, L., van Vliet, G., Boaron, R., Hage, R., de Vos, D. E., Alsters, P. L., Feringa, B. L., *Tetrahedron Lett.*, 2002, 43, 2619.

TABLE 6.5

Green Metrics Summary of *cis*-Cyclooctene to *cis*-Cyclooctene Oxide Reactions Using *t*-Butyl Hydroperoxide as Oxidant[a]

Plan	Catalyst[b]	Recycling Catalyst?	TON	% AE	% Yield	E-Kernel	E-Excess	E-Aux	E-Total	PMI
Sheng[c]	Mo(CO)$_6$	No	1110	63	83	0.91	1.02	0.028	1.96	2.96
Kuhn–Romao[f]	MoO$_2$Cl$_2$ (L1)	No	100	63	100	0.59	0.37	1.41	2.36	3.36
Goncalves[l]	MoCl(L4)(O)$_2$(THF)	No	99	63	99	0.61	0.37	2.06	3.03	4.03
Koner[g]	[Y$_2$(N$_3$)$_2$(nic)$_2$(OH)$_3$(Hnic)(H$_2$O]	Yes	2532	63	85	0.87	0.50	6.99	8.36	9.36
Goncalves[j]	Mesoporous phenylene silica/Cr(CO)$_3$	Yes	10	63	83	0.91	0.51	16.90	18.32	19.32
Neumann[e]	V-MCM41	Yes	8	63	13	11.05	5.42	58.32	74.79	75.79
Zonta–Licini[d]	[Mo]	No	20	63	98	0.63	0	102.45	103.07	104.07
Aziz[k]	Mo(O$_2$)(L3)	Yes	96	63	96	0.65	0	111.83	112.48	113.48
Behzad[m]	VO(L5)	No	971	63	10	15.35	7.36	91.32	114.02	115.02
Machura[h]	ReO(OMe)(L2)$_2$	No	69	63	69	1.30	2.07	133.43	136.80	137.80
Hutchings[i]	Au/graphite	No	0.02	63	11	13.77	6063.56	88.87	6166.20	6167.20

Note: Metrics do not include preparation of catalysts.

[a] Abbreviations and definitions: AE = atom economy, E-kernel = E-factor based on reaction by-products only, E-excess = E-factor based on excess reagent consumption, E-aux = E-factor based on all auxiliary materials used (reaction solvents, catalysts, workup materials, and purification materials), E-total = total E-factor (sum of above three E-factors), NA = not applicable, PMI = process mass intensity (total mass of input materials used divided by mass of target product collected as stated in Jiménez-González, C. et al., *Org. Proc. Res. Dev.*, 2011, 15, 912), TON = turnover number = moles of product/moles of catalyst.

[b] Catalyst abbreviations: Hnic = nicotinic acid, L1 = 2,2-di(1-pyrazolyl)propane, L2 = quinoline-2-carboxylate, L3 = N,N′-bis(2-hydroxybenzylidene)-4,5-dichloro-benzene-1,2-diamine, L4 = 5-methyl-2-(1-methyl-1-trimethylsilanyloxy-ethyl)-1-pyridin-2-yl-cyclohexanol, L5 = N,N′-bis(2-hydroxy-5-bromobenzylidene)-4,5-dinitro-benzene-1,2-diamine, [Mo] = see reference 4, nic = nicotinate, V-MCM = vanadyl silicate mobile crystalline material.

[c] Sheng, M. N., *Synthesis*, 1972, 194.

[d] Romano, F., Linden, A., Mba, M., Zonta, C., Licini, G., *Adv. Synth. Catal.*, 2010, 352, 2937.

[e] Neumann, R., Khenkin, A. M., *Chem. Commun.*, 1996, 2643.

[f] Santos, A. M., Kuhn, F. E., Bruus-Jensen, K., Lucas, I., Romao, C. C., Herdtweck, E., *Dalton Trans.*, 2001, 1332.

[g] Sen, R., Koner, S., Hazra, D. K., Helliwell, M., Mukherjee, M., *Eur. J. Inorg. Chem.*, 2011, 241.

[h] Machura, B., Wolff, M., Tabak, D., Schachner, J. A., Mosch-Zanetti, N. C., *Eur. J. Inorg. Chem.*, 2012, 3764.

[i] Bawaked, S., Dummer, N. F., Dimitratos, N., Bethell, D., He, Q., Kielyb, C. J., Hutchings, G. J., *Green Chem.*, 2009, 11, 1037.

[j] Coelho, A. C., Balula, S. S., Antunes, M. M., Gerganova, T. I., Biona, N., Ferreira, P., Pillinger, M., Valente, A. A., Rocha, J., Goncalves, I. S., *J. Mol. Catal. A* 2010, 332, 13.

[k] Aziz, A. A. A., *J. Mol. Struct.*, 2010, 979, 77.

[l] Valente, A. A., Goncalves, I. S., Lopes, A. D., Rodriguez-Borges, J. E., Pillinger, M., Romao, C. C., Rocha, J., Garca-Merad, X., *New J. Chem.*, 2001, 25, 959.

[m] Rahchamani, J., Behzad, M., Bezaatpour, A., Jahed, V., Dutkiewicz, G., Kubicki, M., Salehi, M., *Polyhedron*, 2011, 30, 2611.

TABLE 6.6
Green Metrics Summary of cis-Cyclooctene to cis-Cyclooctene Oxide Reactions Using Peracids as Oxidants[a]

Plan	Catalyst	Recycling Catalyst?	TON	% AE	% Yield	E-Kernel	E-Excess	E-Aux	E-Total	PMI
Cope[b] (peracetic acid)	NaOAc 3H$_2$O	No	2	66.7	86	0.72	0.34	11.23	12.29	13.29
Reppe[d] (perbenzoic acid)	None	No	NA	60.8	65	2.05	1.19	47.43	50.67	51.67
Harrison[c] (perbenzoic acid resin)	None	No	NA	34.4	95	2.06	2.14	177.90	182.09	183.09
Shea[e] (m-chloro-perbenzoic acid)	None	No	NA	44.6	98	1.29	0.086	138.85	140.23	141.23

Note: Metrics do not include preparation of catalysts.

[a] Abbreviations and definitions: AE = atom economy, E-kernel = E-factor based on reaction by-products only, E-excess = E-factor based on excess reagent consumption, E-aux = E-factor based on all auxiliary materials used (reaction solvents, catalysts, workup materials, and purification materials), E-total = total E-factor (sum of above three E-factors), NA = not applicable, PMI = process mass intensity (total mass of input materials used divided by mass of target product collected as stated in Jiménez-González, C. et al., *Org. Proc. Res. Dev.*, 2011, 15, 912), TON = turnover number = moles of product/moles of catalyst.

[b] Cope, A. C., Fenton, S., Spencer, C. F., *J. Am. Chem. Soc.*, 1952, 74, 5884.

[c] Harrison, C. R., Hodge, P., *Perkin Trans. I*, 1976, 605.

[d] Reppe, W., Schlichting, O., Klager, K., Toepel, T., *Ann. Chem.*, 1948, 560, 1.

[e] Shea, K. J., Kim, J. S., *J. Am. Chem. Soc.*, 1992, 114, 3044.

TABLE 6.7

Green Metrics Summary of cis-Cyclooctene to cis-Cyclooctene Oxide Reactions Using Molecular Oxygen as Oxidant[a]

Plan	Catalyst[b]	Recycling Catalyst?	TON	% AE	% Yield	E-Kernel	E-Excess	E-Aux	E-Total	PMI
Mizuno[c]	$\gamma\text{-SiW}_{10}\{Fe^{3+}(OH_2)\}_2O_{38}^{6-}$	No	9890	100	80	0.25	0.38	1.05	1.67	2.67
Yan[d]	$MgO\text{-}Ti\text{-}N\text{-}SnCl_4$	Yes	41	100	81	0.23	0.55	1.90	2.69	3.69
Rudler[e]	$[Ag\{N(OPPh_2)_2\}]_4$	No	100	100	100	0.00021	0.90	4.17	5.07	6.07
Chandrasekaran[i]	$RuCl_2(L1)_2$	No	40	100	100	0	0.67	44.36	45.03	46.03
Campestrini[h]	$Mo(CO)_6$	No	456	100	38	1.63	0.54	108.99	111.15	112.15
Maldotti[g]	Pd(II)TMPyP Fe(III)TDCPP	Yes	160	100	2	53.39	97.56	82.95	233.90	234.90
Shiraishi[f]	TiO_2	No	0.5	100	29	2.49	0.94	1081.72	1085.15	1086.15

Note: Metrics do not include preparation of catalysts.

a　Abbreviations and definitions: AE = atom economy, E-kernel = E-factor based on reaction by-products only, E-excess = E-factor based on excess reagent consumption, E-aux = E-factor based on all auxiliary materials used (reaction solvents, catalysts, workup materials, and purification materials), E-total = total E-factor (sum of above three E-factors), NA = not applicable, PMI = process mass intensity (total mass of input materials used divided by mass of target product collected as stated in Jiménez-González, C. et al., *Org. Proc. Res. Dev,* 2011, 15, 912), TON = turnover number = moles of product/moles of catalyst.

b　Catalyst abbreviations: L1 = 4,5,4′,5′-tetrahydro-[2,2′]bioxazolyl, TDCPP = meso-tetrakis(2,6-dichlorophenyl) porphyrin, TMPyP = meso-tetrakis(N-methyl-4-pyridyl) porphyrin.

c　Nishiyama, Y., Nakagawa, Y., Mizuno, N., *Angew. Chem. Int. Ed.,* 2001, 40, 3639.

d　Wang, T. J., Ma, Z. H., Yan, Y. Y., Huang, M. Y., Jiang, Y. Y., *Chem. Commun.,* 1996, 1355.

e　Rudler, H., Denise, B., Gregorioa, J. R., Vaissermann, J., *Chem. Commun.,* 1997, 2299.

f　Shiraishi, Y., Morishata, M., Hirai, T., *Chem. Commun.,* 2005, 5977.

g　Maldotti, A., Andreotti, L., Molinari, A., Borisov, S., Vasilev, V., *Chem. Eur. J.,* 2001, 7, 3564.

h　Campestrini, S., Tonellato, U., *Eur. J. Org. Chem.,* 2002, 3827.

i　Kesavan, V., Chandrasekaran, S., *Perkin Trans. I,* 1997, 3115.

TABLE 6.8
Green Metrics Summary of *cis*-Cyclooctene to *cis*-Cyclooctene Oxide Reactions Using Miscellaneous Oxidants[a]

Plan	Oxidant	Catalyst[b]	Recycling Catalyst?	TON	% AE	% Yield	E-kernel	E-Excess	E-Aux	E-Total	PMI
Cope[c]	CrO_3	None	No	NA	87.1	28	3.14	3.33	131.29	137.76	138.76
Ford[f]	Oxone ($KHSO_5$)	None	No	NA	48.1	81	1.58	1.06	232.46	235.10	236.10
Che[e]	Oxone ($KHSO_5$)	$Fe(II)(qpy)Cl_2$	No	17	48.1	58	2.58	2.08	307.36	312.01	313.01
Smith[g]	$PhI = O$	$[Fe(III)T2MpyP]Cl_5$	No	69	38.2	60	3.35	23.12	359.25	385.73	386.73
Mansuy[d]	$PhI = O$	$P_2W_{17}Mn$	No	9	38.2	88	1.97	48.56	386.88	437.40	438.40
Smith[h]	$PhI = O$	Fe(II)TPFPP	Yes	113	38.2	93	1.81	14.69	569.39	585.89	586.89

Note: Metrics do not include preparation of catalysts.

[a] Abbreviations and definitions: AE = atom economy, E-kernel = E-factor based on reaction by-products only, E-excess = E-factor based on excess reagent consumption, E-aux = E-factor based on all auxiliary materials used (reaction solvents, catalysts, workup materials, and purification materials), E-total = total E-factor (sum of above three E-factors), NA = not applicable, PMI = process mass intensity (total mass of input materials used divided by mass of target product collected as stated in Jiménez-González, C. et al., *Org. Proc. Res. Dev.*, 2011, 15, 912), TON = turnover number = moles of product/moles of catalyst.

[b] Catalyst abbreviations: $P_2W_{17}Mn$ = alpha$_2$-[(n-C_4H_9)$_4$N]$_8$P$_2$W$_{17}$O$_{61}$[Mn(III)Br], qpy = 2,2′:6′,2″:6″,2‴-quinquepyridine, T2MpyP = tetrakis(N-methyl-4-pyridyl)porphyrin, TPFPP = tetrakis(pentafluorophenyl)porphyrin.

[c] Cope, A. C., Kinteran, M. R., Keller, R. T., *J. Am. Chem. Soc.*, 1954, 76, 275.

[d] Mansuy, D., Bartoli, J. F., Battioni, P., Lyon, D. K., Finke, R. C., *J. Am. Chem. Soc.*, 1991, 113, 7222.

[e] Chow, T. W. S., Wong, E. L. M., Guo, Z., Liu, Y., Huang, J. S., Che, C. M., *J. Am. Chem. Soc.*, 2010, 132, 13229.

[f] Zhu, W., Ford, W. T., *J. Org. Chem.*, 1991, 56, 7022.

[g] Leanord, D. R., Smith, J. R. L., *Perkin Trans. II*, 1991, 25.

[h] Assis, M. D., Smith, J. R. L., *Perkin Trans. II*, 1998, 2221.

TABLE 6.9
Summary of Overall Green Metrics Rankings according to Best Oxidant and Catalyst Combinations[a]

Plan	Oxidant	Catalyst[b]	Recycling Catalyst?	TON	% AE	% Yield	E-Kernel	E-Excess	E-Aux	E-Total	PMI
Mizuno[c]	O_2	γ-SiW$_{10}${Fe^{3+}(OH$_2$)}$_2$O$_{38}^{6-}$	No	9890	100	80	0.25	0.38	1.05	1.67	2.67
Sheng[d]	t-Butyl hydroperoxide	Mo(CO)$_6$	No	1110	63	83	0.91	1.02	0.028	1.96	2.96
BASF[e]	H_2O_2	Formic acid	No	2	87.5	76	0.50	0.10	4.52	5.13	6.13
Hou (this work, Part 1)	H_2O_2	POM-RTIL (no EtOAc reaction solvent)	Yes	128	87.5	94	0.21	0.0017	7.37	7.59	8.59
Cope[f]	Peracetic acid	NaOAc 3H$_2$O	No	2	66.7	86	0.72	0.34	11.23	12.29	13.29
Hou (this work, Part 1)	H_2O_2	POM-RTIL (EtOAc as reaction solvent)	Yes	125	87.5	90	0.27	0.0017	20.50	20.78	21.78
Hou (this work, Part 2)	H_2O_2	PW$_{12}$	Yes	53	87.5	99	0.15	0.86	38.34	39.36	40.36

Continued

TABLE 6.9 (Continued)
Summary of Overall Green Metrics Rankings according to Best Oxidant and Catalyst Combinations[a]

Plan	Oxidant	Catalyst[b]	Recycling Catalyst?	TON	% AE	% Yield	E-Kernel	E-Excess	E-Aux	E-Total	PMI
Hou (this work, Part 2)	H_2O_2	PW_{11}	Yes	53	87.5	99	0.15	0.86	38.36	39.38	40.38
Hou (this work, Part 2)	H_2O_2	$[DMIM]_5PTiW_{11}O_{40}$	Yes	49	87.5	98	0.17	0.94	40.17	41.28	42.28
Cope[g]	CrO_3	None	No	NA	87.1	28	3.14	3.33	131.29	137.76	138.76
Ford[h]	Oxone ($KHSO_5$)	None	No	NA	48.1	81	1.58	1.06	232.46	235.10	236.10
Smith[i]	$PhI=O$	$[Fe(III)T2MpyP]Cl_5$	No	69	38.2	60	3.35	23.12	359.25	385.73	386.73

[a] Abbreviations and definitions: AE = atom economy, E-kernel = E-factor based on reaction by-products only, E-excess = E-factor based on excess reagent consumption, E-aux = E-factor based on all auxiliary materials used (reaction solvents, catalysts, workup materials, and purification materials), E-total = total E-factor (sum of above three E-factors), NA = not applicable, PMI = process mass intensity (total mass of input materials used divided by mass of target product collected as stated in Jiménez-González, C. et al., *Org. Proc. Res. Dev.*, 2011, 15, 912), TON = turnover number = moles of product/moles of catalyst.

[b] Catalyst abbreviations: C_{12}mim = 3-dodecyl-1-methyl-3H-imidazol-1-ium, POM-RTIL = polyoxometalate room temperature ionic liquid, PW_{11} = $[n-C_{16}H_{33}N(CH_3)_3]_4$ $[Na_3PW_{11}O_{39}]$, T2MpyP = tetrakis(N-methyl-4-pyridyl)porphyrin.

[c] Nishiyama, Y., Nakagawa, Y., Mizuno, N., *Angew. Chem. Int. Ed.*, 2001, 40, 3639.

[d] Sheng, M. N., *Synthesis*, 1972, 194.

[e] Pachaly, H., Schlichting, O., DE 962073 (BASF, 1955).

[f] Cope, A. C., Fenton, S., Spencer, C. F., *J. Am. Chem. Soc.*, 1952, 74, 5884.

[g] Cope, A. C., Kinteran, M. R., Keller, R. T., *J. Am. Chem. Soc.*, 1954, 76, 275.

[h] Zhu, W., Ford, W. T., *J. Org. Chem.*, 1991, 56, 7022.

[i] Leanord, D. R., Smith, J. R. L., *Perkin Trans. II*, 1991, 25.

ACKNOWLEDGMENTS

The authors are grateful for the support from the National Natural Science Foundation of China (No. 21073058), Research Fund for the Doctoral Program of Higher Education of China (20100074110014), and Fundamental Research Funds for the Central Universities, China.

REFERENCES

1. S. A. Nolen, J. Lu, J. S. Brown, P. Pollet, B. C. Eason, K. N. Griffith, R. Gläser, D. Bush, D. R. Lamb, C. L. Liotta, C. A. Eckert. *Ind. Eng. Chem. Res.* 2002, 41, 316–323.
2. W. A. Herrmann, R. M. Kratzer, H. Ding, W. R. Thiel, H. Glas. *J. Organomet. Chem.* 1998, 555(2), 293–295.
3. P. U. Maheswari, X. H. Tang, R. Hage, P. Gamez, J. Reedijk. *J. Mol. Catal. A Chem.* 2006, 258 (1–2), 295–301.
4. M. Guidotti, C. Pirovano, N. Ravasio, B. Lázaro, J. M. Fraile, J. A. Mayoral, B. Coq, A. Galarneau. *Green Chem.* 2009, 11, 1421–1427.
5. Y. Mahha, L. Salles, J. Y. Piquemal, E. Briot, A. Atlamsani, J. M. Brégeault. *J. Catal.* 2007, 249(2), 338–348.
6. K. Kamata, S. Kuzuya, K. Uehara, S. Yamaguchi, N. Mizuno. *Inorg. Chem.* 2007, 46, 3768–3774.
7. K. Kamata, M. Kotani, K. Yamaguchi, S. Hikichi, N. Mizuno. *Chem. Eur. J.* 2007, 13, 639–648.
8. T. Yamase, E. Ishikawa, Y. Asai, S. Kanai. *J. Mol. Catal. A Chem.* 1996, 114(1–3), 237–245.
9. H. Li, Z. S. Hou, Y. X. Qiao, B. Feng, Y. Hu, X. R. Wang, X. G. Zhao. *Catal. Commun.* 2010, 11, 470–475.
10. D. C. Duncan, R. C. Chambers, E. Hecht, C. L. Hill. *J. Am. Chem. Soc.* 1995, 117, 681–691.
11. J. B. Gao, Y. Y. Chen, B. Han, Z. C. Feng, C. Li, N. Zhou, S. Gao, Z. W. Xi. *J. Mol. Catal. A Chem.* 2004, 210(1–2), 197–204.
12. S. S. Wang, W. Liu, Q. X. Wan, Y. Liu. *Green Chem.* 2009, 11(10), 1589–1594.
13. Z. H. Weng, J. Y. Wang, X. G. Jian. *Catal. Commun.* 2008, 9(8), 1688–1691.
14. C. Venturello, E. Alneri, M. Ricci. *J. Org. Chem.* 1983, 48(21), 3831–3833.
15. A. J. Stapleton, M. E. Sloan, N. J. Napper, R. C. Burns. *Dalton Trans.* 2009, 9603–9615.
16. N. V. Maksimchuk, K. A. Kovalenko, S. S. Arzumanov, Y. A. Chesalov, M. S. Melgunov, A. G. Stepanov, V. P. Fedin, O. A. Kholdeeva. *Inorg. Chem.* 2010, 49(6), 2920–2930.
17. Y. Ishii, K. Yamawaki, T. Ura, H. Yamada, T. Yoshida, M. Ogawa. *J. Org. Chem.* 1988, 53(15), 3587–3593.
18. L. Hua, Y. X. Qiao, H. Li, B. Feng, Z. Y. Pan, Y. Y. Yu, W. W. Zhu, Z. S. Hou. *Sci. China Chem.* 2011, 54(5), 769–773.
19. C. Brevard, R. Schimpf, G. Tourne, C. M. Tourne. *J. Am. Chem. Soc.* 1983, 105, 7059–7063.
20. Y. Ishii, K. Yamawaki, T. Ura, H. Yamada, T. Yoshida, M. Ogawa. *J. Org. Chem.* 1988, 53(15), 3587–3593.
21. L. Hua, Y. X. Qiao, Y. Y. Yu, W. W. Zhu, T. Cao, Y. Shi, H. Li, B. Feng, Z. S. Hou. *New J. Chem.* 2011, 35(9), 1836–1841.
22. J. G. Huddleston, A. E. Visser, W. M. Reichert, H. D. Willauer, G. A. Broker, R. D. Rogers. *Green Chem.* 2001, 3(4), 156–164.
23. O. A. Kholdeeva, G. M. Maksimov, R. I. Maksimovskaya, L. A. Kovaleva, M. A. Fedotov, V. A. Grigoriev, C. L. Hill. *Inorg. Chem.* 2000, 39(17), 3828–3837.
24. H. Pachaly, O. Schlichting. DE 962073 (BASF, 1955).

ACKNOWLEDGMENTS

The authors are grateful for the support from the National Natural Science Foundation of China ...

REFERENCES

7 Solvent-Free Flavone Synthesis Reaction Using KHSO$_4$ as Recyclable Catalyst

Ángel G. Sathicq, Diego M. Ruiz, and Gustavo P. Romanelli

CONTENTS

INTRODUCTION

Flavones constitute a large number of natural products with many pharmacological applications. Because of their broad range of biological activities, this class of molecule has been extensively investigated, and more than 10,000 chemically unique flavonoids have been isolated from plants [1]. They exhibit diverse biological and pharmacological activities, for example, anticancer [2], antibacterial, antifungal [3, 4], antioxidant [5], anti-HIV [6], and gastroprotective [7]. Furthermore, it has been reported that some flavones have a repelling property against some phytophagous insects and a subterranean termite (*Coptotermes* sp.) acting as an antifeedant [8].

Various strategies have been developed to synthesize flavones. One of the most commonly used methods consists of the cyclodehydration of 1-(2-hydroxyphenyl)-1,3-diketones. This is usually a catalyzed reaction (Scheme 7.1), and it has been performed in different reaction media. Some reaction conditions employed were the use of excess sulfuric acid [9,10], excess sulfuric acid in glacial acetic acid [11], Co^{III}(salpr)(OH) under neutral condition [12], CuCl$_2$ in ethanol and microwaves [13],

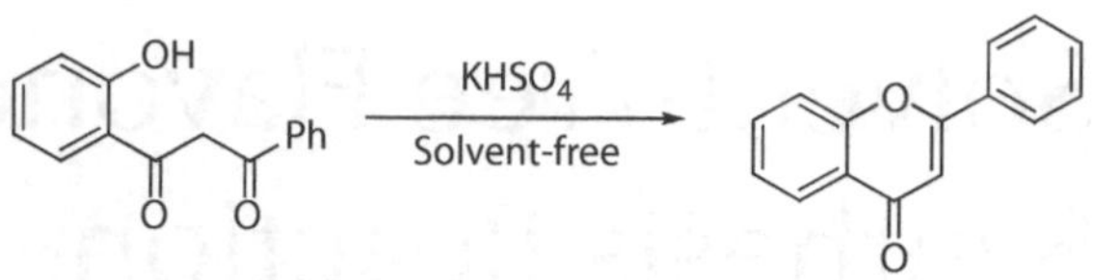

SCHEME 7.1   Flavone synthesis.

P_2O_5 under grinding condition [14], and Preyssler and Wells–Dawson heteropolyacid bulk and supported silica [15, 16].

Other catalytic methods include the cyclization of 3-phenoxycinnamic acid with aluminum chloride [17]; condensation of 2-hydroxyacetophenone with benzoic anhydride using PhCOONa-NaOH [18]; rearrangement of coumarins using HCl [19]; regioselective cyclization of o-alkynoylphenols with DMAP [20], or triflic acid [21]; cyclization of 2-hydroxychalcone using CuI [22], in ionic liquid; $InCl_3$ supported on SiO_2 [23], or SeO_2 [23a], and later the flavanone oxidation; thermal dehydration of 2-hydroxybenzoylacetophenones [24]; direct dehydrogenation of flavanones by tandem bromination [25], using $Tl(OAc)_3$ [26, 27] and $Pd(CF_3CO_2)_2$ [28]; cyclodehydration of 1-(2-bromophenyl)-1,3-diketones [29] or 1-(2-ethoxyphenyl)-1,3-diketones [30], using K_2CO_3 as catalyst; cyclization of 1-phenyl-3-(2-hydroxyphenyl)-2-propenone [31], and acetylene derivatives [32], via flavylium ion; [4+2] oxidative cyclization from 2-hydroxybenzaldehyde and phenylacetylene [33]; one-pot procedure starting from 2-hydroxyacetophenones with 3 equivalents of aroyl chloride in wet K_2CO_3/acetone [34], via ligand-free palladium(II)-catalyzed conjugate addition of arylboronic acids to chromones [35]; tandem reaction via Wittig intermediate [36], from 2-benzoyloxyacetophenone using NaH, oxalic acid, and HCl system [37]; condensation of 2-hydroxyacetophenone and benzoyl chloride with NaOH-nBuN$_4^+$ HSO$_4^-$ under microwave irradiation [38]; and annulation of iodophenol acetates and acetylenes by catalyst of palladium-thiourea-dppp complex [39], via palladium-catalyzed ligand-free cyclocarbonylation of o-iodophenols with terminal acetylenes in phosphonium salt ionic liquids [40] and via Pd-carbene-catalyzed carbonylation reactions of aryl iodides [41]. Figure 7.1 shows various ring construction strategies employed to build the flavone ring.

In this work, we report a simple, clean, and environmentally friendly procedure for the solvent-free preparation of flavone from 1-(2-hydroxyphenyl)-3-phenyl-1,3 propanedione using $KHSO_4$ as recyclable catalyst (Scheme 7.1).

REACTION PROCEDURE

REPRESENTATIVE PROCEDURE FOR THE SYNTHESIS OF FLAVONE USING $KHSO_4$ AS RECYCLABLE CATALYST

A mixture of 24 g (0.1 mol) 1-(2-hydroxyphenyl)-3-phenyl-1,3-propanedione and 40 g potassium bisulfate was heated with stirring for 2 h at 120°C. When the reaction time was over, 120 mL ethyl acetate was added in portions (60 mL and 2 × 30 mL) and the catalyst was filtered. The extracts were combined and the organic solution was concentrated in vacuum giving 98% of crude product. All the solid crude

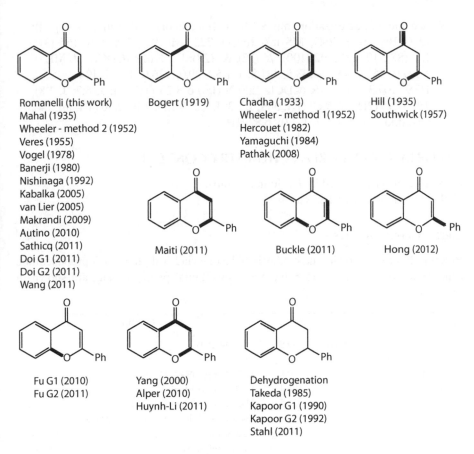

FIGURE 7.1 Ring construction strategies used to construct chromone ring system of flavone via single-step reactions. Bold bonds represent target-forming bonds.

products were recrystallized from methanol (27 mL) with 85% yields (18.9 g); melting point (mp) 97–98°C. Reuse of the recovered dried catalyst gives 95% of crude product under the same conditions.

NOTES

1. The 1-(2-hydroxyphenyl)-3-phenyl-1,3-propanedione (*o*-hydroxydibenzoyl-methane) has been prepared *according to Vogel's Textbook of Practical Organic Chemistry*, 5th edition, p. 1194 [11].
2. The potassium bisulfate was dried in vacuum for 1 h at 100°C.
3. ^{1}H nuclear magnetic resonance (NMR) spectra were recorded on a Varian Mercury Plus 200 spectrometer in CDCl₃, and trimethylsilane (TMS) (0 ppm) was used as the internal reference. ^{13}C NMR was recorded at 50 MHz in CDCl₃, and CDCl₃ (77.0 ppm) was used as the internal reference. A mass spectrum of flavone was recovered in a Micromass LCT spectrometer, and the respective FT-IR spectra in a Bruker IFS 66, at 25°C. The melting point was measured on Buchi–Tottoli apparatus and uncorrected.

4. Flavone characterization: mp, 97–98°C (methanol) (literal mp, 98°C [11]); IR (KBr, cm^{-1}), 1647; MS m/z (int%), 222 (M$^+$, 92), 194 (93), 165 (31), 120 (88), 102 (30), 92 (100), 76 (22), 63 (28); ^{13}C NMR (CDCl$_3$, 50 MHz), δ 177.9, 163.1, 155.9, 133.5, 131.8, 131.5, 129.0, 126.0, 125.5, 124.9, 123.7, 117.9, 107.3; ^{1}H NMR (CDCl$_3$, 200 MHz), δ 8.23 (1H; dd; J, 8.2; J, 1.8), 7.92–7.95 (2H, m), 7.68 (1H; ddd; J, 8.3; J, 7.2; J, 1.7), 7.49–7.56 (4H, m), 7.41 (1H; ddd; J, 8.2; J, 7.2; J, 1.0), 6.82 (1H, s).

IMPORTANCE ON GREEN CHEMISTRY CONTEXT

Data calculations: Offered by Dr. John Andraos.
Yield of the reaction was 85%.

GREEN METRICS ANALYSIS

Below is the calculation for the synthesis of flavone via 1,3-diphenyl-1,3-propanedione (this work), using a new green methodology. Radial pentagon data are shown in Figure 7.2.

Input		Output	
1-(2-Hydroxy-phenyl)-3-phenyl-propane-1,3-dione	24 g	Flavone (pure)	18.9 g
Potassium bisulfate	40 g	Organic solvent waste	129.9 g
Ethyl acetate	108.4 g	Catalyst (recovered)	39 g
Methanol	21.5 g	Unreacted 1-(2-hydroxyphenyl)-3-phenyl-propane-1,3-dione and water by-product	5.1 g
Total	193.9 g	Total waste	168.911 g
		(with catalyst recovery)	135.6 g

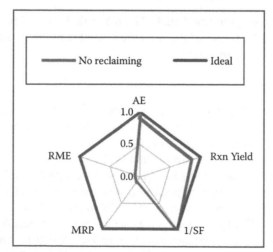

% AE = 92.5
% Yield = 85
E-kernel = 0.27
E-excess = 0
E-aux = 8.97
E-total = 9.24
PMI = 10.24

FIGURE 7.2 Radial pentagon for synthesis of flavone (this work).

E-factor (with catalyst recovery): (174.6 – 39 g waste/18.9 g pure product) = 7.17
E-factor (without catalyst recovery): (174.6 g waste/18.9 g pure product) = 9.24
Mass intensity: (193.9 g raw materials/18.9 g pure product) = 10.24
Atom economy: (222/240) × 100 = 92.5%

CONCLUSION

This catalytic strategy represents an environment-friendly process in comprehensive consideration compared with other catalytic systems listed in publications. The aforementioned method provides a clean, simple, solvent-free reaction and useful alternative for preparing flavone; the use of KHSO$_4$ catalysts provides very good yields, also leading to an easy separation and recovery of the catalysts for further use. The catalytic activity, which was practically constant in consecutive reaction batches, and the high recovery of the catalysts allow for both low environmental impact and low cost. Other green advantages of the method are the low formation of wastes and the replacement of corrosive, soluble mineral acids. This methodology requires the prior preparation of the starting material, unlike other procedures presented in Table 7.1, which are carried out in one-step synthesis through multicomponent reactions [34, 35, 38–41]. However, 1-(2-hydroxy-phenyl)-3-phenyl-propane-1,3-dione is a commercially available material or can be prepared easily through the clean method reported in the literature [42].

EVALUATION FROM THE VIEWPOINT OF THE 12 PRINCIPLES

The catalytic process presented here represents a simple, friendly, safe, and cost-effective route to flavone in an almost quantitative yield and selectivity without utilizing solvent. This is a perfect atom economical reaction. Particularly, the catalyst is inexpensive, thermally stable, recoverable, toxicologically innocuous, and environmentally benign, and has almost negligible vapor pressure. As a consequence, principles 1, 2, 3, 5, 8, 9, and 12 have been highlighted in this procedure.

Principle 1 (prevention): This protocol is clean and no waste is formed during the procedure.

Principle 2 (atom economy): The selected reaction is a high atom-economic reaction, which maximizes the incorporation of all atoms of the substrate into the final product. At the same time, the high yield and selectivity are helpful to realizing the atom economy.

Principle 3 (less hazardous chemical syntheses): Flavone does not represent toxicity to human health and the environment.

Principle 5 (safer solvents and auxiliaries): The solvent-free reaction process reduces the potentially environmental hazards brought about by waste organic solvents.

Principle 8 (reduce derivatives): Derivatization is unnecessary in this procedure.

Principle 9 (catalyst): In particular, KHSO$_4$ is an inexpensive, thermally stable, recoverable, toxicologically innocuous, and environmentally benign catalyst for this transformation, and it has almost negligible vapor pressure.

TABLE 7.1

Green Metrics Summary of 35 One-Step Reactions to Flavone[a]

Plan	Green Technology	Synthesis Strategy	% AE	% Yield	E-Kernel	E-Excess	E-Aux	E-Total	PMI
Romanelli [this work]	Solvent-free reaction	[6+0] condensation	92.5	85b,c	0.27	0	8.97	9.24	10.24
Alper [40]	MCR; ionic liquid reaction solvent	[3+2+1] cyclization	63.2	95	0.66	6.05	7.15	13.86	14.86
Bogert [17]		[6+0] Friedel–Crafts cyclization	51.9	85	1.26	3.59	13.33	18.19	19.19
Huynh and Li [41]	MCR	[3+2+1] cyclization	52.5	98	0.94	15.00	8.68	24.62	25.62
Makrandi [14]	Mechanochemistry	[6+0] condensation	58.1	80	1.15	0	26.81	27.96	28.96
Hercouet [36]		Tandem acylation–[5+1] cyclization–elimination	36.1	96	1.89	0.54	28.68	31.11	32.11
Van Lier [23]		[6+0] oxidative cyclization	92.5	96	0.13	0.39	30.54	32.06	33.06
Pathak [38]	Mechanochemistry; microwaves	[5+1] cyclization	66.8	81	0.85	0.47	36.91	38.23	39.23
Kabalka [13]	Microwaves	[6+0] condensation	92.5	98	0.11	0	41.75	41.86	42.86
Autino [15]	Solvent-free reaction	[6+0] condensation	92.5	89	0.21	0	47.64	47.85	48.85
Mahal [43]		[6+0] oxidative cyclization	79.4	44	1.88	1.74	44.47	48.08	49.08
Southwick [32]		[6+0] cyclization via flavylium ion intermediate	64.3	54	1.87	15.15	37.36	54.38	55.38
Wang [22]	Ionic liquid solvent	[6+0] oxidative cyclization	92.5	96	0.13	2.27	56.52	58.92	59.92
Yamaguchi [37]		[5+1] condensation via rearrangement	62.7	83	1.25	0.47	65.27	66.66	67.66
Stahl [28]		Dehydrogenation	92.5	88	0.22	0.56	75.28	76.06	77.06

Sathicq [16]	Solvent-free reaction	[6+0] condensation	92.5	87	0.25	0	89.54	89.79	90.79
Doi G2 [21]	100% atom economical	[6+0] cyclization	100	80	0.25	0	99.55	99.80	100.80
Wheeler (method 2) [24]		[6+0] condensation	92.5	94	0.14	0	128.5	128.64	129.64
Doi G1 [20]	100% atom economical	[6+0] cyclization	100	96	0.041	0	192.53	192.57	193.57
Kapoor G1 [26]		Dehydrogenation	36.7	96	1.84	0.18	192.39	194.42	195.42
Yang [39]	MCR	[3+2+1] cyclization	36.0	92	2.02	20.81	175.11	197.95	198.95
Veres [19]		Rearrangement	83.5	72	0.67	0	205.28	205.95	206.95
Vogel [11]		[6+0] condensation	92.5	95	0.14	0	230.36	230.51	231.51
Kapoor G2 [27]		Dehydrogenation	36.7	96	1.84	0.18	250.23	252.26	253.26
Wheeler (method 1) [9]		[5+1] condensation via rearrangement	92.5	43	1.50	0	314.33	315.33	316.33
Nishinaga [12]		[6+0] condensation	92.5	86	0.26	0	351.43	351.69	352.69
Maiti [33]		[4+2] oxidative cyclization	92.5	80	0.35	0.76	377.51	378.62	379.62
Takeda [25]		Tandem bromination–dehydrobromination	30.8	97	2.34	0.20	433.38	435.91	436.91
Chadha [18]		Tandem acylation–[5+1] cyclization	61.3	9	16.75	55.59	377.27	449.60	450.60
Hill [31]		[6+0] cyclization via flavylium ion intermediate	73.9	27	3.97	14.83	444.22	463.03	464.03
Fu G1 [29]		[6+0] cyclization	50.4	91	1.19	0	519.14	520.32	521.32
Fu G2 [30]		[6+0] condensation	82.8	85	0.42	0	536.15	536.57	537.57
Buckle [34]		[4+2] oxidative cyclization	53.6	60	2.09	11.99	578.28	592.37	593.37

Continued

TABLE 7.1 (*Continued*)
Green Metrics Summary of 35 One-Step Reactions to Flavone[a]

Plan	Green Technology	Synthesis Strategy	% AE	% Yield	E-Kernel	E-Excess	E-Aux	E-Total	PMI
Banerji [10]		[6+0] condensation	92.5	84	0.29	0	780.87	781.16	782.16
Hong [35]		Tandem direct carbon–carbon coupling–dehydrogenation	37.2	94	1.86	6.44	4171.13	4179.43	4180.43

Abbreviations and definitions: AE = atom economy, E-kernel = E-factor based on reaction by-products only, E-excess = E-factor based on excess reagent consumption, E-aux = E-factor based on all auxiliary materials used (reaction solvents, catalysts, workup materials, and purification materials), MCR = multicomponent reaction, E-total = total E-factor (sum of above three E-factors), PMI = process mass intensity (total mass of input materials used divided by mass of target product collected as stated in Jiménez-González, C. et al., *Org. Proc. Res. Dev.*, 2011, 15, 912).

Note: The algorithms used to calculate metrics have been published [44–46].

[a] Entry in bold text corresponds to present submission.

[b] Crude product yield was 98%.

[c] First and second reuses of the catalyst were 98% and 97%, respectively.

Principle 12 (inherently safer chemistry for accident prevention): Substances and the form of a substance used in this protocol do not represent substances for potential chemical accidents.

REFERENCES

1. G. Agati, E. Azzarello, S. Pollastri, M. Tattini, *Plant Sci.*, 2012, 196, 67–76.
2. S. Martens, A. Mithöfer, *Phytochemistry*, 2005, 66, 2399–2407.
3. S. Alam, *J. Chem. Sci.*, 2004, 116, 325–331.
4. H. Göker, D. Boykin, S. Yildiz, *Bioorg. Med. Chem.*, 2005, 13, 1707–1714.
5. H. Chu, H. Wu, Y. Lee, *Tetrahedron*, 2004, 60, 2647–2655.
6. J. Wu, X. Wang, Y. Yi, K. Lee, *Bioorg. Med. Chem. Lett.*, 2003, 13, 1813–1815.
7. J. Ares, P. Outt, J. Randall, P. Murray, P. Weisshaar, L. Obrien, B. Ems, S. Kakodkar, G. Kelm, W. Kershaw, K. Werchowski, A. Parkinson, *J. Med. Chem.*, 1995, 38, 4937–4943.
8. M. Morimoto, K. Tanimoto, S. Nakano, T. Ozaki, *J. Agric. Food Chem.*, 2003, 51, 389–393.
9. T.S. Wheeler, *Org. Synth.*, 1952, 32, 72–76 (method 2).
10. A. Banerji, N.C. Goomer, *Synthesis*, 1980, 874–875.
11. B.S. Furniss, A.J. Hannaford, V. Rogers, P.W.G. Smith, A.R. Tatchell, *Vogel's Textbook of Practical Organic Chemistry*, 5th ed., Longman, New York, 1989, p. 1194.
12. A. Nishinaga, H. Ando, K. Maruyama, T. Machino, *Synthesis*, 1992, 839–841.
13. G. Kabalka, A. Mereddy, *Tetrahedron Lett.*, 2005, 46, 6315–6317.
14. D. Sharma, J.K. Makrandi, *Green Chem. Lett. Rev.*, 2009, 2, 157–159.
15. G.P. Romanelli, E.G. Virla, P.R. Duchowicz, A.L. Gaddi, D.M. Ruiz, D.O. Bennardi, E.V. Ortiz, J.C. Autino, *J. Agric. Food Chem.*, 2010, 58, 6290–6295.
16. D.O. Bennardi, G.P. Romanelli, A.G. Sathicq, J.C. Autino, G.T. Baronetti, H.J. Thomas, *Appl. Catal. A*, 2011, 404, 68–73.
17. M.T. Bogert, J.K. Marcus, *J. Am. Chem. Soc.*, 1919, 41, 83–107.
18. T.C. Chadha, K. Venkataraman, *J. Chem. Soc.*, 1933, 1073–1076.
19. K. Veres, V. Horak, *Coll. Czech. Chem. Commun.*, 1955, 20, 371–373.
20. M. Yoshida, Y. Fujino, K. Saito, T. Doi, *Tetrahedron*, 2011, 67, 9993–9997 (generation 1, G1).
21. M. Yoshida, Y. Fujino, T. Doi, *Org. Lett.*, 2011, 13, 4526–4529 (generation 2, G2).
22. Z. Du, H. Ng, K. Zhang, H. Zeng, J. Wang, *Org. Biomol. Chem.*, 2011, 9, 6930–6933.
23. N. Ahmed, H. Ali, J.E. van Lier, *Tetrahedron Lett.*, 2005, 46, 253–256.
24. T.S. Wheeler, *Org. Synth.*, 1952, 32, 72–76 (method 1).
25. N. Takeno, T. Fukushima, S. Takeda, K. Kishimoto, *Bull. Chem. Soc. Jpn.*, 1985, 58, 1599–1600.
26. O.V. Singh, R.P. Kapoor, *Tetrahedron Lett.*, 1990, 31, 1459–1462 (generation 1, G1).
27. M.S. Khanna, O.V. Singh, C.P. Garg, R.P. Kapoor, *Perkin Trans. I*, 1992, 2565–2568 (generation 2, G2).
28. T. Diao, S.S. Stahl, *J. Am. Chem. Soc.*, 2011, 133, 14566–14569.
29. J. Zhao, Y. Zhao, H. Fu, *Angew. Chem. Int. Ed.*, 2011, 50, 3769–3773 (generation 1, G1).
30. J. Zhao, Y. Zhao, H. Fu, *Org. Lett.*, 2012, 14, 2710–2713 (generation 2, G2).
31. D.W. Hill, R.R. Melhuish, *J. Chem. Soc.*, 1935, 1161–1166.
32. P.L. Southwick, J.R. Kirchner, *J. Am. Chem. Soc.*, 1957, 79, 689–691.
33. G. Maiti, R. Karmakar, R.N. Bhattacharya, U. Kayal, *Tetrahedron Lett.*, 2011, 52, 5610–5612.
34. C.F. Chee, M.J.C. Buckle, N.A. Rahman, *Tetrahedron Lett.*, 2011, 52, 3120–3123.
35. D. Kim, K. Ham, S. Hong, *Org. Biomol. Chem.*, 2012, 10, 7305–7312.
36. A. Hercouet, M. Le Corre, *Synthesis*, 1982, 597–598.

37. I. Hirao, M. Yamaguchi, M. Hamada, *Synthesis*, 1984, 1076–1078.
38. V.N. Pathak, R. Gupta, B. Varshney, *J. Heterocyclic Chem.*, 2008, 45, 589–592.
39. H. Miao, Z. Yang, *Org. Lett.*, 2000, 2, 1765–1768.
40. Q. Yang, H. Alper, *J. Org. Chem.*, 2010, 75, 948–950.
41. L. Xue, L. Shi, Y. Han, C. Xia, H.V. Huynh, F. Li, *Dalton Trans.*, 2011, 7632–7638.
42. D. Sharma, S. Surender Kumar, J.K. Makrandi, *Green Chem. Lett. Rev.*, 2009, 2, 53–59.
43. H. S. Mahal, H. S. Rai, K. Venkataranman, *J. Chem. Soc.*, 1935, 866–868.
44. J. Andraos, *Org. Process Res. Dev.*, 2009, 13, 161–185.
45. J. Andraos, M. Sayed, *J. Chem. Educ.* 2007, 84, 1004–1010.
46. J. Andraos, *The Algebra of Organic Synthesis: Green Metrics, Design Strategy, Route Selection, and Optimization*, CRC Press, Boca Raton, FL, 2012.

8 Waste-Minimized Synthesis of γ-Nitroketones

Eleonora Ballerini, Ferdinando Pizzo, and Luigi Vaccaro

CONTENTS

INTRODUCTION

Aliphatic nitrocompounds have proven to be valuable intermediates mainly due to the electron-withdrawing property of the nitro group as well as the possibility of its straightforward conversion into several other useful functional groups. In fact, the literature constantly reports advances about their preparation or utilization.[1] The conjugate addition of nitroalkanes to α,β-unsaturated compounds is a widely studied process for the formation of new carbon-carbon bonds and for the preparation of γ-nitro carbonyl compounds and γ-nitro esters.[1,2] Recent contributions aimed at the improvement of this process have been focused on the definition of novel or asymmetric catalytic systems as well as the use of equimolar amounts of reactants.[3,4] Recently, we have focused our work on the optimization of novel synthetic tools by employing eco-friendly reaction protocols based on the adoption of green reaction media such as water[5] or solvent-free conditions (SolFCs),[6] combined with the use of heterogeneous catalysts.[7] Our optimization approach has been finalized toward waste minimization (E-factor minimization)[8] by exploiting flow chemistry to set single- and multistep procedures.[7c,e,f,h] Flow approach

111

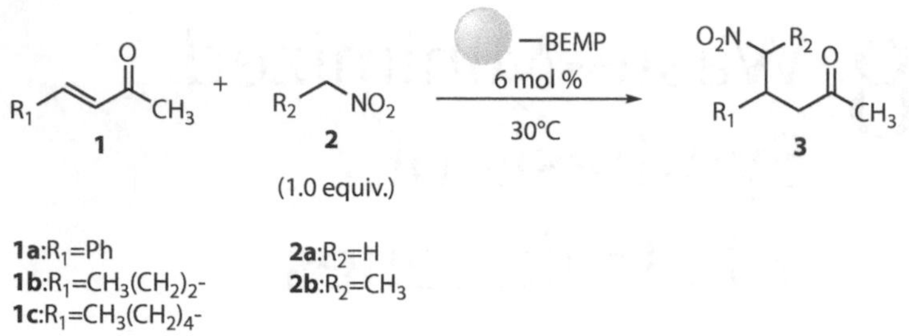

SCHEME 8.1 PS-BEMP-catalyzed Michael addition of nitroalkanes **2** to α,β-enones **1** in a cyclic flow reactor.

can be very efficient to reduce the E-factor,[8] and in general the environmental cost of synthetic processes. Highlights of our results are the recovery of the products with a minimal amount of organic solvent, as well as the reproducible recovery of the solid catalyst and its repetitive use in consecutive processes. We have studied the reactivity of nitro compounds in unconventional media, and as a continuation of our interest in this context,[9,10] we have defined an efficient protocol for the Michael addition of nitroalkanes (**2a,b**) to α,β-unsaturated ketones (α,β-enones, **1a–c**), catalyzed by 2-tert-butylimino-2-diethylamino-1,3-dimethylperhydro-1,3,2-diazaphosphorine supported on polystyrene (PS-BEMP)[7c] in flow (Scheme 8.1).

REACTION PROCEDURES

REPRESENTATIVE PROCEDURE FOR THE NUCLEOPHILIC ADDITION OF NITROALKANES 2A,B TO α,β-UNSATURATED KETONES 1A–C CATALYZED BY PS-BEMP IN A CYCLIC FLOW REACTOR

A premixed equimolar mixture of (E)-benzylidene acetone (**1a**) (note 1, 100 mmol, 14.62 g) and nitroethane (**2b**) (note 2, 100 mmol, 7.18 mL) was charged into a glass column (note 3) functioning as a reservoir, and then ethyl acetate (note 4, 0.04 mL/mmol of enone **1a**, 4 mL) was added. PS-BEMP (note 5, 6 mol%, 2.72 g), suitably dispersed in 1 mm diameter solid glass beads (note 6), was charged on a glass column (note 7); the equipment was installed into a thermostated box and connected, by using the appropriate valves, to a pump. The schematic representation of the reactor is depicted in Figure 8.1 (thermostated box is not shown for clarity). The reaction mixture was continuously pumped (note 8, flow rate 1.5 mL/min) through the catalyst column at 30°C for 4.0 h to reach the complete conversion of (E)-benzylidene acetone (**1a**) to 5-nitro-4-phenylhexan-2-one (**3a**) (note 9). At this stage, the pump was left to run in order to recover the reaction mixture into a flask. To completely recover the product and clean the reactor, ethyl acetate (4 × 4 mL, 0.16 mL/mmol of enone **1a**) was pumped through the system (each portion for 10 min) and then collected into the flask. Then the solvent was removed by vacuum distillation (note 10) to isolate the product **3a** in a syn/anti 1/1 diastereoisomeric mixture and in 98% yield. With this procedure the catalyst was left inside the reactor and was successfully reused for two other subsequent runs,

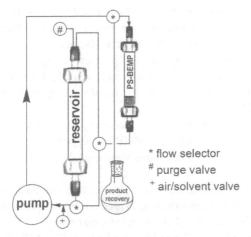

* flow selector
# purge valve
+ air/solvent valve

FIGURE 8.1 Flow reactor scheme using PS-BEMP as catalyst for the Michael addition of nitroethane (**2b**) to (E)-benzylidene acetone (**1a**).

achieving identical results. All products **3a–c** (note 11) were identified by ^{1}H nuclear magnetic resonance (^{1}HNMR), ^{13}C NMR, and gas chromatography electron impact mass spectrometry (GC-EIMS).

NOTES

1. (E)-Benzylidene acetone (**1a**) (99%) was purchased from Sigma-Aldrich and used without further purification.
2. Nitroethane (**2b**) (≥98%) was purchased from Sigma-Aldrich and used without further purification.
3. Glass columns and the corresponding screwed caps were purchased from Sigma-Aldrich (Omnifit Glass Column L × i.d. 150 × 25 mm).
4. Ethyl acetate (99.5%) was purchased from Sigma-Aldrich and used as received. 0.04 mL of ethyl acetate/mmol of enone **1a** was added for lowering the viscosity of the reaction product. In order to generalize this protocol and to allow an easier reproducibility for nonexperts, this operation was done for all the substrates even when unnecessary. It is important to notice that the amount of EtOAc has been optimized to conserve the advantage of an increased and necessary reactivity of the catalytic system under SolFCs and achieve a lower viscosity of the reaction mixture.[7e] Higher dilution results in lower or very poor reaction conversions.[7f]
5. PS-BEMP (200–400 mesh, 2.2 mmol BEMP/g) was purchased from Sigma-Aldrich and dried for 24 h before use. We routinely confirm the catalyst loading by elemental analyses (elemental analyses were realized by using a FISONS instrument EA 1108 CHN), and we have found that loading of commercial PS-BEMP may vary. Therefore, we have decided to use 6 mol% instead of 5 mol% of PS-BEMP, as previously reported,[7c] because it guarantees more confident results.

6. Glass beads were purchased from Sigma-Aldrich and their use to disperse the solid catalyst is particularly advised to avoid malfunctioning of the pump when the reaction product is very viscous.
7. Glass columns and the corresponding screwed caps were purchased from Sigma-Aldrich (Omnifit Glass Column L × i.d. 150 × 15 mm).
8. The pump used was a chemically resistant diaphragm metering pump, STEPDOS 03S.
9. The progress of the reaction was controlled by GC analysis that was performed by using Hewlett-Packard HP 5890A equipped with a capillary column DB-35MS (30 m, 0.53 mm), a flame ionization detector (FID) and hydrogen as the gas carrier.
10. The distillation temperature was 45°C; the distillate flask and the vapor trap were cooled with an ice-salt bath to recover 19 mL (95%) of solvent.
11. Detailed NMR and GC-EIMS data information:

    5-Nitro-4-phenylhexan-2-one (**3a**) (98% yield) ^{1}H NMR (400 MHz, CDCl$_3$) diastereoisomer A δ 1.32 (d, 3H, J = 6.6 Hz), 2.01 (s, 3H), 2.74 (dd, 1H, J = 17.0, 4.3 Hz), 2.97 (dd, 1H, J = 17.0, 9.6 Hz), 3.66–3.77 (m, 1H), 4.71–4.81 (m, 1H), 7.18 (s, 1H), 7.20 (d, 1H, J = 1.5 Hz), 7.23–7.38 (m, 3H); diastereoisomer B δ 1.48 (d, 3H, J = 6.7 Hz), 2.12 (s, 3H), 2.90 (dd, 1H, J = 17.5, 7.6 Hz), 3.05 (dd, 1H, J = 17.5, 6.7), 3.66–3.77 (m, 1H), 4.88 (q, 1H, J = 6.6 Hz), 7.13 (d, 1H, J = 1.4 Hz), 7.15 (d, 1H, J = 1.7 Hz), 7.23–7.38 (m, 3H); ^{13}C NMR (100.6 MHz, CDCl$_3$) diastereoisomer A δ 17.6, 30.3, 45.2, 46.1, 87.0, 127.8, 128.1, 128.9, 138.2, 204.9; diastereoisomer B δ 16.6, 30.4, 44.5, 44.6, 85.8, 127.8, 128.0, 128.6, 137.8, 205.6; GC-EIMS (m/z, %) diastereoisomer A 221 (M$^+$, n.d.), 174 (32), 43 (100); diastereoisomer B 221 (M$^+$, n.d.), 174 (33), 43 (100).

    4-(Nitromethyl)heptan-2-one (**3b**) (95% yield) elemental analysis calculated: C, 55.47; H, 8.73; N, 8.09; found: C, 55.43; H, 8.77; N, 8.02; Fourier transform infrared (FT-IR) (CHCl$_3$) 1371, 1552, 1715 cm^{-1}; ^{1}H NMR (400 MHz, CDCl$_3$) δ 0.85–0.95 (m, 3H), 1.35–1.45 (m, 4H), 2.17 (s, 3H), 2.50–2.65 (m, 3H), 4.45 (d, 2H, J = 5.4 Hz); ^{13}C NMR (100.6 MHz, CDCl$_3$) δ 13.8, 19.7, 30.3, 32.7, 33.5, 44.5, 78.3, 206.6; GC-EIMS (m/z, %) 173 (M$^+$, n.d.), 43 (73), 55 (92), 69 (100).

    4-(Nitromethyl)nonan-2-one (**3c**) (94% yield) elemental analysis calculated: C, 59.68; H, 9.52; N, 6.96; found: C, 59.63; H, 9.50; N, 6.92; FT-IR (CHCl$_3$) 1372, 1551, 1716 cm^{-1}; ^{1}H NMR (400 MHz, CDCl$_3$) δ 0.85–0.89 (m, 3H), 1.27–1.35 (m, 8H), 2.17 (s, 3H), 2.50–2.65 (m, 3H), 4.43 (d, 2H, J = 5.4 Hz); ^{13}C NMR (100.6 MHz, CDCl3) δ 13.8, 22.3, 26.2, 30.4, 31.3, 31.5, 32.9, 44.5, 78.3, 206.6; GC-EIMS (m/z, %) 201 (M$^+$, n.d.), 69 (34), 58 (43), 55 (52), 43 (100).

DISCUSSION

The wide range of applications of PS-BEMP as catalyst in a cyclic flow was further confirmed by carrying out the reactions of nitromethane (**2a**) or nitroethane (**2b**) with three different α,β-unsaturated ketones, **1a–c**. The obtained results are reported in Table 8.1.

TABLE 8.1
PS-BEMP (6 mol%) Catalyzed Michael Addition of Nitroalkanes 2 to α,β-Enones 1 at 30°C and in Flow

Entry	α,β-Enone	Nitroalkane	t (h)	Product	Yield (%)[a]
1	1a	2b	4	3a	98[b]
2	1b	2a	4	3b	95
3	1c	2a	4	3c	94

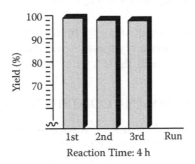

[a] Isolated yield of the pure product **3**.
[b] The product **3a** was obtained in a syn/anti 1/1 diasteroisomeric mixture.

FIGURE 8.2 Reuse of PS-BEMP in the Michael addition of nitroethane (**2b**) to (E)-benzylidene acetone (**1a**) in flow.

At 30°C and in the presence of 6 mol% of PS-BEMP, enones **1a–c** reacted with nitroalkanes **2a, b** to give the desired products in 4 h. In the case of **1a** and **2b** the desired product **3a** was obtained as a syn/anti 1/1 diasteroisomeric mixture in 98% yield (Table 8.1, entry 1). Alkyl-substituted α,β-enones **1b, c** reacted efficiently with nitromethane **2a** in the presence of 6 mol% of PS-BEMP and gave the corresponding γ-nitro ketones **3b, c** in satisfactory yields (95% and 94% respectively, Table 8.1, entries 2 and 3). In the case of the reaction of **1a** and **2b** the catalyst was successfully reused for three consequent runs (see Figure 8.2) and in 4 h reaction time product **3b** was always obtained in 95% to 98% yields.

To keep the highest efficiency at the fourth run, reactivation of PS-BEMP was necessary, and this can be performed directly by flowing a 0.13 M ethyl acetate solution of BEMP (1.2 equivalents) through the catalyst glass column. After regeneration, the catalyst was reused for three more runs, showing the same efficiency as in the first run.

IMPORTANCE IN GREEN CHEMISTRY CONTEXT

In order to evaluate the efficiency of our synthetic route to γ-nitroketones and compare it to those of already published protocols, we have calculated the most relevant green metrics performances according to the previously described Andraos algorithm, which has been used to test the greenness of six industrial and nine academic plans to some target molecules.[11] The metrics examined are percent overall yield, percent overall atom economy (% AE),[12] reaction mass efficiency (RME),[12] overall E-factor,[13] and process mass intensity (PMI).[14] The overall E-factor in turn was subdivided into its components arising from by-products, side products, and unreacted starting materials (E-kernel), excess reagent consumption (E-excess), and auxiliary material consumption arising from reaction solvent, catalysts, all workup materials, and all purification materials (E-aux).

Data calculations: Offered by Dr. John Andraos.

As shown in Tables 8.2 and 8.3, our synthetic method is the best compared to the literature reports, and the E-factor is lower than the others, even when it has been calculated without considering the recycling of the catalyst and the solvent recovery (Figure 8.3).

In comparison with other methods of synthesis of 5-nitro-4-phenylhexan-2-one (3a), our synthetic approach certainly prevails over them in terms of efficiency and sustainability (see Scheme 8.2 and Table 8.4).

HIGHLIGHTS OF THE PRESENT CONTRIBUTION

Our protocol represents an environment-friendly process and a waste-minimized method for synthesizing γ-nitroketones with high yields; this synthetic tool maximizes the incorporation of all materials used in the process into the final product, therefore resulting in a perfect atom economical method. Our approach based on the use of a reusable heterogeneous catalyst, solvent-free conditions, equimolar amounts of reagents, and minimal use of organic solvent to recover the final γ-nitrocarbonyl compounds product allowed us to reduce dramatically the reaction's waste, thus minimizing the E-factor. The solid catalyst was successfully recovered and reused with no loss of efficiency for three consecutive runs. The minimized amount of organic solvent has been used to recover the γ-nitrocarbonyl compounds and clean the reactor system. Ethyl acetate has been chosen as a desirable green option also included in the Pfizer[24] and other[25] green solvents lists. This solvent is generally safe and possesses an adequate boiling point value that has allowed it to be almost fully recovered. In conclusion, the combination of flow conditions in a large scale with recoverable heterogeneous catalysts represents a very promising approach able to drastically reduce the environmental cost of organic synthesis.

TABLE 8.2
Summary of Metrics of Literature Examples according to the Synthetic Strategy Shown in Scheme 8.1

Catalyst[a]	% AE	% Yield	RME	E-Kernel	E-Excess	E-Aux	E-Total	PMI
PS-BEMP[b] (flow with catalyst and solvent recovery)	100	98	98	0.02	0	0.042	0.066	1.07
PS-BEMP[b] (flow with no catalyst or solvent recovery)	100	98	98	0.02	0	0.96	0.98	1.98
PS-BEMP[7c] (batch)	100	98	98	0.020	0.0012	5.10	5.12	6.12
PS-TBD[7a]	100	95	93.4	0.052	0.019	5.51	5.58	6.58
NaOMe[15,c]	66	49	41.1	2.07	0.90	3.63	6.61	7.61
[bmim][BF$_4$]/ K$_2$CO$_3$[16]	100	82	72.2	0.22	0.17	102.03	102.42	103.42
DBU[17]	100	70	65.6	0.40	0.096	141.77	142.26	143.26
Rasta-TBD[6d]	100	84	84	0.19	0.0030	146.77	146.96	147.96
Hydrotalcite/μW[18]	100	89	89	0.12	0	229.44	229.56	230.56
KO$_2$/18-crown-6-ether[19]	100	70	52.3	1.01	28.77	409.05	438.84	439.84

[a] Abbreviations: [bmim][BF$_4$] = 1-butyl-3-methylimidazolium tetrafluoroborate, DBU = 1,8-diazabicyclo[5,4,0]-undecene, PS-TBD = polystyryl-supported 1,5,7-triazabicyclo[4,4,0]dec-5-ene, μW = microwaves.
[b] Our synthetic method.
[c] The lower atom economy is due to a subsequent neutralization of sodium methoxide with acetic acid.

TABLE 8.3
Summary of Metrics Showing Comparison with Enantioselective Syntheses

Catalyst[a,b]	% AE	% Yield	RME	E-Kernel	E-Excess	E-Aux[c]	E-Total	PMI
[(S,S)-(Salen) Al]$_2$O/Et$_3$N[3c]	100	90	84	0.11	1.51	6.90	8.52	9.52
BMIC[3d,20]	100	80	7.9	0.26	11.45	0.25	11.95	12.95
ACBTPT[21]	100	82	11	0.22	6.22	10.27	16.71	17.71
(S)-5-Pyrrolidin-2-yl-1H-tetrazole[3e,f]	100	67	50	0.49	0.51	1687.4	1688.4	1689.4

[a] Abbreviations: BMIC = 4S-Benzyl-1-methyl-imidazolidine-2R-carboxylic acid, ACBTPT = (S,S)-1-(2-aminocyclohexyl)-3-[3,5-bis(trifluoromethylphenyl)]-thiourea.
[b] These calculations are shown separately in this table to distinguish the difference between racemic and stereoselective synthesis.
[c] In these examples, column chromatography purification is always necessary, and it has not been included in the metrics calculation.

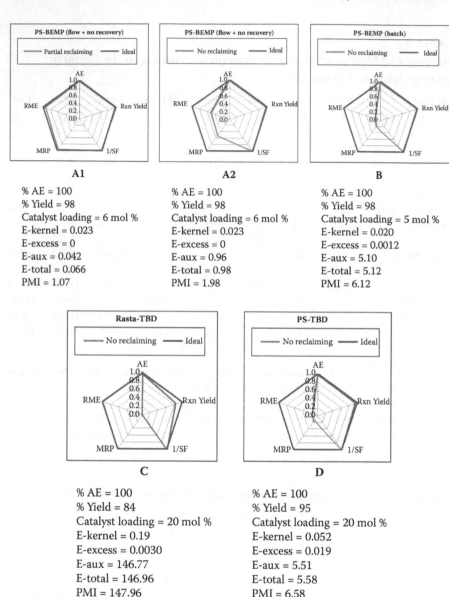

FIGURE 8.3 Radial pentagons showing metric performances for reactions shown in Table 8.2 under various conditions: (A1) PS-BEMP catalyst under continuous flow procedure with recovery of catalyst and 95% of ethyl acetate solvent (this work), (A2) PS-BEMP catalyst under continuous flow procedure with no recovery of catalyst or solvent, (B) PS-BEMP catalyst under batch procedure (see reference 7c), (C) Rasta-TBD catalyst (see reference 6d), and (D) PS-TBD catalyst (see reference 7a).

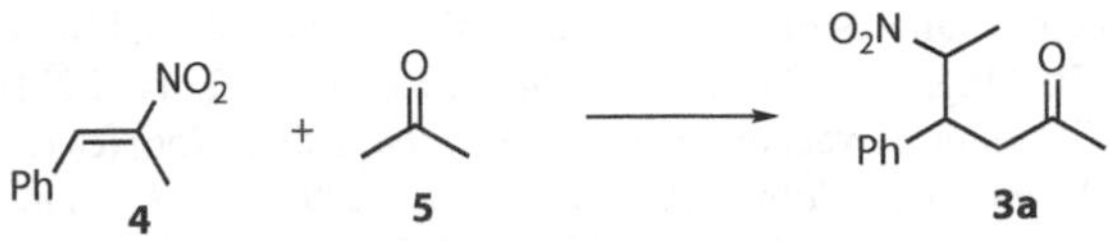

SCHEME 8.2 Synthesis of 5-nitro-4-phenylhexan-2-one (**3a**) from trans-β-methyl-β-nitrostyrene (**4**) and acetone (**5**).

TABLE 8.4
Summary of Metrics of Literature Examples according to the Synthesis Strategy Shown in Scheme 8.2

Catalyst[a]	% AE	% Yield	E-Kernel	E-Excess	E-Aux[b]	E-Total	PMI
Chiral DPT[22]	100	98	0.02	0.53	6.86	7.41	8.41
PMM[23]	100	97	0.031	2.43	478.70	481.16	482.16

[a] Abbreviations: DPT = diphenylphosphinothioamide, PMM = 4-pyrrolidin-2S-yl-methyl-morpholine.
[b] The purification step has not been included in the calculation of the E-factor.

ACKNOWLEDGMENTS

We gratefully acknowledge the Ministero dell'Istruzione, dell'Università e della Ricerca (MIUR) for the project "Firb-Futuro in Ricerca" and the Università degli Studi di Perugia for financial support.

REFERENCES

1. (a) D. Seebach, E. W. Colvin, F. Lehr, T. Weller, *Chimia* 1979, 33, 1–18; (b) M. Yamaguchi, T. Shiraishi, Y. Igarashi, M. Hirama, *Tetrahedron Lett.* 1994, 35, 8233; (c) M. Yamaguchi, Y. Igarashi, R. S. Reddy, T. Shiraishi, M. Hirama, *Tetrahedron* 1997, 53, 11223; (d) R. Ballini, in *Studies in Natural Products Chemistry*, vol. 19, ed. Atta-ur-Rahman, Elsevier, Amsterdam, 1997, p. 117; (e) R. Ballini, *Synlett* 1999, 1009; (f) S. Hanessian, V. Pham, *Org. Lett.* 2000, 2, 2975; (g) S. C. Bergmeier, *Tetrahedron* 2000, 56, 2561; (h) N. Ono, *The Nitro Group in Organic Synthesis*, John Wiley: New York, 2001; (i) R. Ballini, G. Bosica, D. Fiorini, A. Palmieri, M. Petrini, *Chem. Rev.* 2005, 105, 933.
2. (a) G. Rosini, R. Ballini, *Synthesis* 1988, 833; (b) *Nitro Compounds: Recent Advances in Synthesis and Chemistry*, ed. H. Feuer, A. T. Nielsen, VCH, Weinheim, 1990.
3. Asymmetric version: for reviews concerning enantioselective conjugate additions see: (a) P.I. Dalko, L. Moisan, *Angew Chem.* 2004, 116, 5248; *Angew. Chem. Int. Ed.* 2004, 43, 5138; for some recent literature examples see: (b) S. Hanessian, S. Govindan, J. S. Warrier, *Chirality* 2005, 17, 540; (c) M. S. Taylor, D. N. Zala-tan, A. M. Lerchner, E. N. Jacobsen, *J. Am. Chem. Soc.* 2005, 127, 1313; (d) A. Prieto, N. Halland, K. A. Jorgensen, *Org. Lett.* 2005, 7, 3897; (e) C. E. T. Mitchell, S. E. Brenner, S. V. Ley, *Chem. Commun.* 2005, 5346; (f) C. E. T. Mitchell, S. E. Brenner, J. Garcia-Fortanet, S. V. Ley, *Org. Biomol. Chem.* 2006, 4, 2039; (g) S. B. Tsogoeva, S. B. Jagtap, Z. A. Ardemasova, *Tetrahedron Asymm.* 2006, 17, 989; (h) W. Ye, D. Leow, S. Goh, M. L. C.-T. Tan, C.-H. Chian, *Tetrahedron Lett.* 2006, 47, 1007.

4. For some recent examples of the racemic version see: (a) R. Ballini, G. Bosica, D. Livi, A. Palmieri, R. Maggi, G. Sartori, *Tetrahedron Lett.* 2003, 44, 2271; (b) F. Fringuelli, F. Pizzo, C. Vittoriani, L. Vaccaro, *Chem. Commum.* 2004, 2756; (c) R. Ballini, D. Fiorini, M. V. Gil, A. Palmieri, *Tetrahedron* 2004, 60, 2799; (d) R. Ballini, C. Balsamini, G. Diamantini, N. Savoretti, *Synthesis* 2005, 1055; (e) Y. Liang, D. Dong, Y. Lu, Y. Wang, W. Pan, Y. Chai Q. Liu, *Synthesis* 2006, 3301; (f) S. Gabrielli, A. Palmieri, A. Perosa, M. Selva, R. Ballini, *Green Chem.* 2011, 13, 2026; (g) R. Mancuso, A. Palmieri, R. Ballini, B. Gabriele, *Tetrahedron* 2012, 68, 5852.

5. As representative examples see: (a) S. Bonollo, D. Lanari, F. Pizzo, L. Vaccaro, *Org. Lett.* 2011, 13, 2150; (b) S.Bonollo, D. Lanari, L. Vaccaro, *Eur. J. Org. Chem.* 2011, 2587; (c) S. Bonollo, D. Lanari, A. Marrocchi, L. Vaccaro, *Curr. Org. Synth.* 2011, 8, 319.

6. As representative examples see: (a) F. Fringuelli, R. Girotti, F. Pizzo, L. Vaccaro, *Org. Lett.* 2006, 8, 2487; (b) L. Castrica, F. Fringuelli, F. Pizzo, L. Vaccaro, *Lett. Org. Chem.* 2008, 5, 602; (c) D. Lanari, R. Ballini, A. Palmieri, F. Pizzo, L. Vaccaro, *Eur. J. Org. Chem.* 2011, 2874; (d) S. Bonollo, D. Lanari, T. Angelini, F. Pizzo, A. Marrocchi, L. Vaccaro, *J. Catal,* 2012, 285, 216.

7. (a) F. Fringuelli, F. Pizzo, C. Vittoriani, L. Vaccaro, *Chem. Commun.* 2004, 2756; (b) F. Fringuelli, F. Pizzo, C. Vittoriani, L. Vaccaro, *Eur. J. Org. Chem.* 2006, 1231; (c) R. Ballini, L. Barboni, L. Castrica, F. Fringuelli, D. Lanari, F. Pizzo, L. Vaccaro, *Adv. Synth. Catal.* 2008, 350, 1218; (d) F. Fringuelli, D. Lanari, F. Pizzo, L. Vaccaro, *Eur. J. Org. Chem.* 2008, 3928; (e) A. Zvagulis, S. Bonollo, D. Lanari, F. Pizzo, L. Vaccaro, *Adv. Synth. Catal.* 2010, 352, 2489; (f) F. Fringuelli, D. Lanari, F. Pizzo, L. Vaccaro, *Green Chem.* 2010, 12, 1301; (g) T. Angelini, F. Fringuelli, D. Lanari, L. Vaccaro, *Tetrahedron Lett.* 2010, 51, 1566; (h) D. Lanari, R. Ballini, S. Bonollo, A. Palmieri, F. Pizzo, L. Vaccaro, *Green Chem.* 2011, 13, 3181; (i) D. Lanari, F. Montanari, F. Marmottini, O. Piermatti, M. Orrù, L. Vaccaro, *J. Catal.* 2011, 277, 80; (l) S. Calogero, D. Lanari, M. Orrù, O. Piermatti, F. Pizzo, L. Vaccaro. *J. Catal.* 2011, 282, 112.

8. (a) R. A. Sheldon, *Chem. Ind.* (London), 1997, 12; (b) R. A. Sheldon, *Green Chem.* 2007, 9, 1273; (c) J. Augé, *Green Chem.* 2008, 10, 225; (d) R. A. Sheldon, *Chem. Commun.* 2008, 3352.

9. (a) F. Fringuelli, F. Pizzo, S. Tortoioli, L. Vaccaro, *Org. Lett.* 2005, 7, 4411; (b) F. Fringuelli, R. Girotti, O. Piermatti, F. Pizzo, L. Vaccaro, *Org. Lett.* 2006, 8, 5741; (c) R. Ballini, L. Barboni, F. Fringuelli, A. Palmieri, F. Pizzo, L. Vaccaro, *Green Chem.* 2007, 9, 823.

10. (a) M. Baumann, I. R. Baxendale, S. V. Ley, *Mol. Diversity* 2011, 15, 613; (b) C. J. Smith, C. D. Smith, N. Nikbin, S. V. Ley, I. R. Baxendale, *Org. Biomol. Chem.* 2011, 9, 1927; (c) M. O'Brien, N. Taylor, A. Polyzos, I. R. Baxendale, *Chem. Sci.* 2011, 2, 1250; (d) C. Aranda, A. Cornejo, J. M. Fraile, E. García-Verdugo, M. J. Gil, S. V. Luis, J. A. Mayoral, V. Martinez-Merino, Z. Ochoa, *Green Chem.* 2011, 13, 983; (e) V. Sans, N. Karbass, M. I. Burguete, V. Compañ, E. García-Verdugo, S. V. Luis, M. Pawlak, *Chem. Eur. J.* 2011, 17, 1894; (f) M. Baumann, I. R. Baxendale, M. Brasholz, J. J. Hayward, S. V. Ley, N. Nikbin, *Synlett* 2011, 1375; (g) L. J. Martin, A. L. Marzinzik, S. V. Ley, I. R. Baxendale, *Org. Lett.* 2011, 13, 320.

11. J. Andraos, *Org. Process Res. Dev.* 2009, 13, 161.

12. B. M. Trost, *Science* 1991, 254, 1471.

13. R. A. Sheldon, *ChemTech* 1994, 24(3), 38.

14. C. Jiménez-Gonzàlez, C. S. Ponder, Q. B. Broxterman, J. B. Manley, *Org. Process Res. Dev.* 2011, 15, 912.

15. L. I. Smith, J. S. Showell, *J. Org. Chem.* 1952, 17, 827.

16. S. G. Zlotin, A. V. Bogolyubov, G. V. Kryshtal, G. M. Zhdankina, M. I. Struchkova, V. A. Tartakovsky, *Synthesis* 2006, 22, 3849.

17. Y. Matsuzawa, K. Ichimura, K. Kudo, *Inorg. Chim. Acta* 1998, 277, 151.

18. S. Vijaikumar, K. Pitchumani, *Ind. J. Chem.* 2010, 49B, 469.
19. K. N. Singh, R. S. Raghuvanshi, M. Singh, *Ind. J. Chem.* 2007, 46B, 829.
20. N. Halland, R. G. Hazell, K. A. Jorgensen, *J. Org. Chem.* 2002, 67, 8331.
21. K. Mei, M. Jin, S. Zhang, P. Li, W. Liu, X. Chen, F. Xue, W. Duan, W. Wang, *Org. Lett.* 2009, 11, 2864.
22. A. Lu, T. Liu, R. Wu, Y. Wang, G. Wu, Z. Zhou, J. Fang, C. Tang, *J. Org. Chem.* 2011, 76, 3872.
23. M. Betancourt, K. Sakthivel, R. Thayumanavan, F. Tanaka, C. F. Barbas III, *Synthesis*, 2004, 1509.
24. K. Alfonsi, J. Colberg, P. J. Dunn, T. Fevig, S. Jennings, T. A. Johnson, H. P. Kleine, C. Knight, M. A. Nagy, D. A. Perry, M. Stefaniakc, *Green Chem.* 2008, 10, 31.
25. C. Capello, U. Fischer, K. Hungerbühler, *Green Chem.* 2007, 9, 927.

9 Green Syntheses of Ethers and Esters Using Dimethyl Carbonate

Amarajothi Dhakshinamoorthy,
Kuppusamy Kanagaraj, and Kasi Pitchumani

CONTENTS

INTRODUCTION

Alkylation of phenol (etherification) or carboxylic acid (esterification), an important reaction in organic chemistry, is widely used in the synthesis of petrochemicals, fine chemicals, and pharmaceuticals [1, 2]. Methyl halides, dimethyl sulfate, and diazomethane are the commonly used methylating reagents in conventional organic synthesis [3], but they result in large amounts of environmentally hazardous by-products like hydrohalides and sulfur dioxide, and in the case of diazomethane, also involve safety issues due to its explosive nature. In addition, most of the reported procedures for esterification/etherification require use of sulfuric acid, hydrochloric acid, and other environmentally hazardous chemicals as catalysts/reagents. Thus, there is always scope to develop an environmentally benign catalytic system that can promote the above-mentioned transformations with high yield of the products without compromising the total cost of the process.

In the past two decades, dimethyl carbonate (DMC) has emerged as a nontoxic green alternative methylating reagent [4] to replace dimethyl sulfate or methyl halide, and a number of methods have been reported for DMC-mediated processes [5, 6]. For example, methylation of 2-naphthol has been reported using potassium carbonate as catalyst in the presence of tetramethylammonium chloride as methylating reagent with 87% yield [7]. Quantitative yield of 1,6-dimethoxynaphthalene is recovered when the reaction is performed using sodium hydroxide as catalyst using dimethyl sulfate in ethanol [8]. On the other hand, DMC has been used as a green methylating reagent with a wide range of catalysts such as fluorine-containing metal oxides [9], potassium carbonate [10], ionic liquids [11], potassium carbonate combined with PEG 1000 [12], Hβ [13], NaX [14], 1,8-diazabicyclo[5,4,0]undec-7-ene (DBU) [15, 16], KBr/SiO$_2$ [17], tetrabutylammonium bromide as phase transfer catalyst [18], tetrabutylammonium bromide as catalyst under semicontinuous process [19], calcined Mg-Al hydrotalcites [20], and C,N-chelated organotin(IV) compounds [21]. NaX has also been used as catalyst for the methylation of phenol using DMC in the presence of dimethyl sulfoxide (DMSO) as solvent [22]. However, in most of the cases, DMC was used as solvent and reagent, which reduces the overall cost and workup procedure. The green metric analysis of LDH-Met is compared with the reported catalytic systems in methylation of phenol (Table 9.1).

Esterification of carboxylic acids has been reported using 1,2-dimethylimidazole [23], sulfuric acid [24], DBU under different reaction conditions [25–27], and Bu$_3$N [28]. Although these homogeneous catalysts result in high yields of the desired products, recovery, reuse, and isolation of the products pose tedious complications. In order to overcome these limitations, Hβ and HZSM-5 zeolites [29] are used as catalysts for the esterification of benzoic acid. The green metric analysis of LDH-Met is compared with the reported catalytic systems in methylation of benzoic acid (Table 9.2).

The method discussed here reports the catalytic activity of L-methionine supported on layered double hydroxide (LDH-Met) as catalyst in *O*-methylation of phenols (etherification) and esterification of carboxylic acids under mild reaction conditions in the absence of any solvent. Specific details of this work can be found in the communication we published in *Chem. Eur. J.* [30]. Scheme 9.1 represents LDH-Met catalyzed methylation reactions for phenol and benzoic acid, respectively.

REPRESENTATIVE PROCEDURE FOR O-METHYLATION OF PHENOLS

LDH-Met (100 mg) was charged in an autoclave bomb followed by the addition of 0.1 mL (1 mmol) of phenol. To this slurry, 1.2 mL (13 mmol) of DMC was added. This heterogeneous mixture was placed inside a preheated oven at 180°C. The reaction was monitored by analyzing the reaction mixture at different time intervals by gas chromatography. The absence of phenol was noticed at 6 h. Then, the autoclave was removed from an oven, cooled to room temperature. To this heterogeneous mixture, ethyl acetate (5 mL) was added and extracted the products for 6 h. This solution was filtered, and the solid catalyst was recovered and dried for the next run. The filtrate was washed with water (5 mL) and the organic layer was dried over 250 mg of anhydrous sodium sulfate. Products were isolated and were confirmed by their ^{1}H nuclear magnetic resonance (NMR) spectra (adapted from reference [30]).

TABLE 9.1

Comparison of Green Metrics Analysis for Methylation of Phenol with Reported Catalytic Systems

Catalyst	Conditions	% AE	% Yield	E-Kernel	E-Excess	E-Aux	E-Total	PMI
Bu$_4$N$^+$Br[18]	DMC, K$_2$CO$_3$, 5 h, 90–100°C	62.07	99	763.70	0	2.75	766.45	767.45
Bu$_4$N$^+$Br[19]	3.5 h, 130°C using semicontinuous process	62.07	100	1.52	115.73	120.30	237.55	238.55
DBU[15]	DMF, 220°C, 10 min in continuous microwave-heated reactor	62.07	96	154.48	0.31	2.98	157.77	158.77
K$_2$CO$_3$	DMF, 150°C, 4.5 h, triglyme (triethylene glycol dimethyl ether)	62.07	97	0.25	6.67	95.32	102.23	103.23
Calcined Mg-Al hydrotalcites[20]	240–275°C in vapor phase in a fixed-bed down-flow silica reactor	62.07	95	77.65	0	5.20	82.85	83.85
LDH-Met[30] (present study)	**DMC, 6 h, 180°C in autoclave**	**62.07**	**92**	**0.09**	**0**	**54.01**	**54.10**	**55.10**
K$_2$CO$_3$[10]	5 h, 150°C	62.07	70	42.83	0.36	0.63	43.82	44.82
C,N-chelated organotin(IV) compounds[21]	6 h, 200°C in autoclave	62.07	94	0.78	0	37.87	38.65	39.65
[BMIm]Cl[11]	120°C, 1.5 h at atmospheric pressure	62.07	100	7.60	0	2.23	9.83	10.83
HTs-F[9]	8 h, 200°C in autoclave	62.07	99	7.95	0	1.56	9.52	10.52
KBr/SiO$_2$[17]	8 h, 200°C in autoclave	62.07	95	7.76	0	1.68	9.45	10.45
DBU[26]	DMF, 2.0 min heated using microwave	62.07	90	0.40	0	4.88	5.28	6.28

Abbreviations: DBU = 1,8-diazabicyclo[5.4.0]undec-7-ene, [BMIm]Cl = 1-n-butyl-3-methylimidazolium chloride, HTs-F = fluorine-modified Mg-Al mixed oxides, PTC = phase transfer catalyst.

TABLE 9.2
Comparison of Green Metrics Analysis for Methylation of Benzoic Acid with Reported Catalytic Systems

Catalyst	Conditions	% AE	% Yield	E-Kernel	E-Excess	E-Aux	E-Total	PMI
DMI[23]	DMF, 12 min, 120°C in microwave	64.15	99	0.063	8.61	121.69	130.36	131.36
DBU[26]	DMF, 45 s, heated	64.15	97	0.39	46.34	47.74	94.47	95.47
Bu₃N[28]	DMF, 285°C, 150 bar in microwave	64.15	86	0.17	40.88	48.57	89.61	90.61
DBU[25]	MeCN, 12 min in microwave	64.15	99	0.63	35.61	46.13	82.36	83.36
DBU[27]	DMF, 2.0 min, 90°C in microwave	64.15	97	0.64	26.79	36.88	64.30	65.30
LDH-Met[30] (present study)	**DMC, 6 h, 180°C in autoclave**	**64.15**	**89**	**0.07**	**0**	**57.55**	**57.62**	**58.62**
Zeolites Hβ and HZSM5[29]	4 h, 200°C, 1 atm	64.15	98	4.11	1.11	4.90	10.13	11.13
H₂SO₄[24]	8 h, 80–85°C	64.15	94	3.26	1.05	2.85	7.16	8.16

Abbreviations: DMI = 1,2-Dimethylimidazole, DBU = 1,8-diazabicyclo[5,4,0]undec-7-ene.

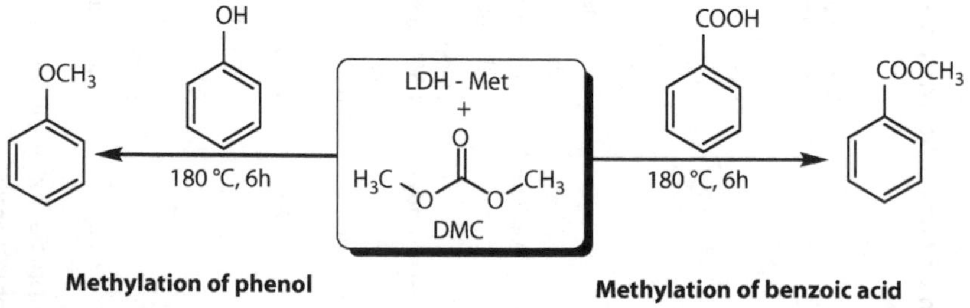

Methylation of phenol   **Methylation of benzoic acid**

SCHEME 9.1 Methylation of phenol and benzoic acid catalyzed by LDH-Met with DMC.

REPRESENTATIVE PROCEDURE FOR ESTERIFICATION OF CARBOXYLIC ACIDS

LDH-Met (100 mg) was charged in an autoclave bomb followed by the addition of 0.1 g (0.8 mmol) of benzoic acid. To this slurry, 1.2 mL (13 mmol) of DMC was added. This heterogeneous mixture was placed inside a preheated oven at 180°C. The reaction was monitored by analyzing the reaction mixture at different time intervals by gas chromatography. The absence of benzoic acid was noticed at 6 h. Then, the autoclave was removed from an oven, and cooled to room temperature. To this heterogeneous mixture, ethyl acetate (5 mL) was added and extracted the products for 6 h. This solution was filtered, and the solid catalyst was recovered and dried for the next run. The filtrate was washed with water (5 mL), and the organic layer was dried over 250 mg of anhydrous sodium sulfate. Products were isolated and were confirmed by their ^{1}H-NMR spectra (adapted from reference [30]).

PREPARATION OF LAYERED DOUBLE HYDROXIDE

Aluminum (III) nitrate nonahydrate (0.01 M) and magnesium (II) nitrate hexahydrate (0.05 M) were dissolved in deionized water (100 mL) and added slowly to a second solution (60 mL) of sodium carbonate (0.03 M) and sodium hydroxide (0.07 M). The resulting mixture was heated at 65°C for 18 h with vigorous stirring. The white slurry was then cooled to room temperature, filtered, washed well with deionized water (200 mL), and dried in a hot air oven at 110°C (adapted from reference [30]).

PREPARATION OF L-METHIONINE INTERCALATED LAYERED DOUBLE HYDROXIDE

A sodium salt of L-methionine (1 g) in 5 mL of water was added to 2 g of calcined layered double hydroxide (activated at 250°C for 3 h in oxygen atmosphere), and the resulting mixture was stirred for 24 h at room temperature in nitrogen atmosphere. The white slurry was then filtered and dried in vacuum (adapted from reference [30]).

NOTES

1. Phenol and benzoic acid and their derivatives were obtained from Sigma Aldrich and used without further purification.
2. Dimethyl carbonate was also purchased from Sigma Aldrich and used as received.
3. L-Methionine was obtained from Sigma Aldrich.
4. Ethyl acetate was purchased from Merck and used as received.
5. Aluminum nitrate, magnesium nitrate, sodium carbonate, and sodium hydroxide were obtained from Merck with analytical grade quality.
6. High-purity (99.99%) gases were used for the experiments.
7. Autoclave bomb was designed at parent institution with the following specifications: 25 ml of Teflon container with cap surrounded by a stainless steel container with a screw-type cap.

8. Temperature-controlled oven was used to maintain the temperature exactly as per the requirement.

9. The percentage conversion, the purity, and relative yields of the final products were confirmed and their characterizations were carried out using gas chromatograph (Shimadzu GC-17A model, SE-30 (10%) capillary column with flame ionization detector (FID) detector and high-purity nitrogen as carrier gas).

10. Proton Nuclear Magnetic Resonance (^{1}H-NMR) spectra were recorded on BRUKER 300 MHz using trimethylsilane (TMS) as internal standard in CDCl$_3$. The following ^{1}H-NMR data were adapted from reference [30].

11. Anisole: δ (ppm) 6.7–7.3 (m, 5H), 3.9 (s, 3H).

12. 2-Chloroanisole: δ (ppm) 6.6–7.6 (m, 4H), 3.8 (s, 3H).

13. 4-Bromoanisole: δ (ppm) 7.3 (d, J= 8.7 Hz, 2H), 6.7 (d, J= 8.7 Hz, 2H), 3.7 (s, 3H).

14. 2-Methoxynaphthalene: δ (ppm) 7.1–7.9 (m, 7H), 3.9 (s, 3H).

15. Methyl phenyl sulfide: δ (ppm) 7.0–7.6 (m, 5H), 2.5 (s, 3H).

16. Methyl benzoate: δ (ppm) 7.4–7.80 (m, 5H), 3.9 (s, 3H).

17. Methyl 4-methoxybenzoate: δ (ppm) 7.9 (d, $J = 9.0$ Hz, 2H), 6.9 (d, $J = 9.0$ Hz, 2H), 3.88 (s, 3H), 3.85 (s, 3H).

18. Methyl 3-nitrobenzoate: δ (ppm) 8.8 (t, $J = 2.1$ Hz, 1H), 8.3–8.4 (m, 2H), 7.6 (t, $J = 7.8$ Hz, 1H), 3.8 (s, 3H).

19. Methyl indole-5-carboxylate: δ (ppm) 8.6 (s, 1H), 8.4 (s, 1H), 7.9 (d, $J = 8.7$ Hz, 1H), 7.3 (d, $J = 8.4$ Hz, 1H), 7.2 (dd, $J = 4.5, 1.8$ Hz, 1H), 6.6 (s, 1H), 3.9 (s, 3H).

20. Methyl 2-phenylacetate: δ (ppm) 7.2 (m, 5H), 3.7 (s, 3H), 3.6 (s, 2H).

21. Methyl cinnamate: δ (ppm) 7.7–7.2 (m, 5H), 6.4 (dd, $J = 12.9, 3.0$ Hz, 2H), 3.8 (s, 3H).

GREEN METRICS ANALYSIS

Calculations for the synthesis of anisole from phenol and methyl benzoate from benzoic acid (this article) are shown below.

Methylation of Phenol				Methylation of Benzoic Acid			
Input		Output		Input		Output	
Phenol	0.107 g	Anisole	0.113 g	Benzoic acid	0.1 g	Methyl benzoate	0.099 g
DMC	1.284 g	Organic solvent waste	4.485 g	DMC	1.284 g	Organic solvent waste	4.485 g
LDH-Met (catalyst)	0.1 g	Recovered catalyst	0.1 g	LDH-Met	0.1 g	Recovered catalyst	0.1 g
Ethyl acetate	4.485 g	Sodium sulfate	0.25 g	Ethyl acetate	4.485 g	Sodium sulfate	0.25 g
Total (used)	**5.976 g**	Total (waste)	5.763	Total (used)	**5.969 g**	Total (waste)	5.77

METHYLATION OF PHENOL

E-factor (for reaction only) = 11.31
E-factor for complete reaction processes = 54.10
Process mass intensity (PMI) = 55.10
Atom economy = 62.07%
Reaction yield = 92%

METHYLATION OF BENZOIC ACID

E-factor (for reaction only) = 12.98
E-factor for complete reaction processes = 57.62
Process mass intensity (PMI) = 58.62
Atom economy = 64.15%
Reaction yield = 89%

Abbreviations and definitions:

AE = atom economy.
E-kernel = E-factor based on reaction by-products only.
E-excess = E-factor based on excess reagent consumption.
E-aux = E-factor based on all auxiliary materials used (reaction solvents, catalysts, workup materials, and purification materials).
E-total = total E-factor (sum of above three E-factors).
PMI = 1 + E-total.

DISCUSSION

The scope of LDH-Met catalyst is also extended to substituted phenols and the representative examples are given in Table 9.3, but more results can be found in our

TABLE 9.3
LDH-Met-Catalyzed Chemoselective
O-Methylation of Phenols Using Dimethyl
Carbonate as Methylating Agent at 6 h

Run	Substrate	Product	Yield (%)
1	Phenol	Anisole	92
2	2-Chlorophenol	2-Chloroanisole	90
3	4-Bromophenol	4-Bromoanisole	95
4	4-Nitrophenol	4-Nitroanisole	89
5	2-Naphthol	2-Methoxynaphthalene	94
6	Thiophenol	Methyl phenyl sulfide	97

Source: Adapted from Dhakshinamoorthy, A., Sharmila, A., and Pitchumani, K., *Chem. Eur. J.*, 16, 1128, 2010 [30].

earlier communication [30]. In most of the cases, the yields of the products were more than 90%, suggesting better activity of this catalyst with various substrates.

Recently, Selva and coworkers reported zeolite-based chemoselective esterification of indolecarboxylic acids with DMC as methylating agent [31]. These observations have prompted us to extend the scope of this protocol for esterification of various carboxylic acids under identical conditions as employed for phenol. It has been observed that many carboxylic acids are alkylated in good yields, but selected examples from our earlier communication [30] are shown in Table 9.4. Benzoic acids with electron donating and withdrawing substituents were alkylated in high yields. Further, indole-5-carboxylic acid was alkylated in 89% yield under identical reaction conditions.

To ascertain the catalyst stability, the recovered catalyst from the first run was subjected to consecutive runs, and the observed results are given in Table 9.5. The catalyst can be reused three times without any change in the conversion and selectivity of the product.

TABLE 9.4
LDH-Met-Catalyzed Chemoselective Esterification of Carboxylic Acids Using Dimethyl Carbonate as Methylating Agent at 6 h

Run	Substrate	Product	Yield (%)
1	Benzoic acid	Methyl benzoate	89
2	4-Methoxybenzoic acid	Methyl 4-methoxybenzoate	88
3	3-Nitrobenzoic acid	Methyl 3-nitrobenzoate	89
4	Indole-5-carboxylic acid	Methyl indole-5-carboxylate	89
5	2-Phenylacetic acid	Methyl 2-phenylacetate	84
6	Cinnamic acid	Methyl cinnamate	84

Source: Adapted from Dhakshinamoorthy, A., Sharmila, A., and Pitchumani, K., *Chem. Eur. J.*, 16, 1128, 2010 [30].

TABLE 9.5
Recycling Test for the O-Methylation of Phenol Catalyzed by LDH-Met Using DMC

Cycle	1	2	3	4
Selectivity (%)	100	100	100	100
Yield (%)	90	91	88	89

Source: Adapted from Dhakshinamoorthy, A., Sharmila, A., and Pitchumani, K., *Chem. Eur. J.*, 16, 1128, 2010 [30].

CONCLUSIONS

In conclusion, the selective O-methylation of phenols with DMC has been reported in the presence of LDH-Met as an efficient organocatalyst representing an environmentally benign process. Phenols and carboxylic acids are converted quantitatively into their corresponding aryl methyl ethers and aryl methyl esters, respectively, in high selectivities and yields, and no C-methylation products are obtained. Separation of unreacted substrates is not required. The products are obtained in high purity and do not need further purification. LDH-Met can be easily recovered and can be recycled without any change in its activity.

EVALUATION FROM THE VIEWPOINT OF THE 12 PRINCIPLES

The catalytic system reported here represents a simple, environmentally benign, and cost-effective method for the synthesis of ethers and esters in quantitative yields with very high selectivity. This procedure demonstrates a one-step green protocol for the synthesis of ethers and esters. The catalyst is thermally stable, recoverable, and reusable, uses environmentally benign reagents, and generates no hazardous by-products. Green chemistry principles 1–5, 8, 9, and 12 have been highlighted using this catalytic system.

Principle 1 (prevention): This protocol is clean and no toxic waste is formed during the course of the reaction.

Principle 2 (atom economy): This reaction is significantly atom economical as the generated by-products, methanol and carbon dioxide, can be potentially recycled back to DMC by reported routes.

Principle 3 (less hazardous syntheses): No rigorous reaction conditions are involved in this protocol. Nontoxic by-products, namely, methanol and carbon dioxide, are formed, unlike in previous reported procedures, which involve environmentally hazardous reagents and unacceptable products.

Principle 4 (design benign chemicals): The prepared esters and ethers are not very toxic to human health and the environment.

Principle 5 (safer solvents and auxiliaries): This method does not use any solvent.

Principle 8 (reduce derivatives): Derivatization is not required in this procedure.

Principle 9 (catalyst): This catalyst is thermally stable, recoverable, reusable, and environmentally benign for this transformation.

Principle 12 (inherently safer chemistry for accident prevention): Substances and reagents used in this procedure do not have any potential risk toward chemical accidents.

ACKNOWLEDGMENTS

KP thanks Department of Science and Technology (DST), New Delhi, India, and AD thanks University Grants Commission (UGC), New Delhi, for the award of assistant professorship under its Faculty Recharge Programme.

REFERENCES

1. P. H. Gore, G. A. Olah (eds.), Friedel–Crafts and Related Reactions, vol. III, Wiley-Interscience, New York, 1964.
2. J. Cejka, B. Wichterlova, *Catal. Rev. Sci. Eng.* 2002, 44, 375.
3. A. G. Shaikh, S. Sivaram, *Chem. Rev.* 1996, 96, 951.
4. P. Tundo, M. Selva, *Acc. Chem. Res.* 2002, 35, 706.
5. G. Barcelo, D. Grenouillat, J.-P. Senet, G. Sennyey, *Tetrahedron* 1990, 46, 1839.
6. W. C. Shieh, S. Dell, O. Repic, *Org. Lett.* 2001, 3, 4279.
7. N. Maras, S. Polanc, M. Kocevar, *Tetrahedron* 2008, 64, 11618.
8. T. Zhang, Q. Yang, H. Shi, L. Chi, *Org. Proc. Res. Dev.* 2009, 13, 647.
9. G. Wu, X. Wang, B. Chen, J. Li, N. Zhao, W. Wei, Y. Sun, *Appl. Catal. A Gen.* 2007, 329, 106.
10. S. Ouk, S. Thiebaud, E. Borredon, P. Le Gars, *Green Chem.* 2002, 4, 431.
11. Z. L. Shen, X. Z. Jiang, W. M. Mo, B. X. Hu, N. Sun, *Green Chem.* 2005, 7, 97.
12. A. Bomben, M. Selva, P. Tundo, *Ind. Eng. Chem. Res.* 1999, 38, 2075.
13. S. R. Kirumakki, N. Nagaraju, K. V. R. Chary, S. Narayanan, *J. Catal.* 2004, 221, 549.
14. M. Selva, E. Militello, M. Fabris, *Green Chem.* 2008, 10, 73.
15. U. Tilstam, *Org. Process Res. Dev.* 2012, 16, 1150.
16. F. Rajabi, M. R. Saidi, *Synth. Commun.* 2004, 34, 4179.
17. B. Chen, Q. N. Dong, N. Zhao, W. Wei, Y. H. Sun, *J. Chem. Soc. Pak.* 2009, 31, 552.
18. S. Ouk, S. Thiebaud, E. Borredon, P. Legars, L. Lecomte, *Tetrahedron Lett.* 2002, 43, 2661.
19. S. Ouk, S. Thiebaud, E. Borredon, P. Le Gars, *Appl. Catal. A Gen.* 2003, 241, 227.
20. T. M. Jyothi, T. Raja, M. B. Talawar, K. Sreekumar, S. Sugunan, B. S. Rao, *Synth. Commun.* 2000, 30, 3929.
21. T. Weidlich, L. Dusek, B. Vystrcilova, A. Eisner, P. Svec, A. Ruzicka, *Appl. Organometal. Chem.* 2012, 26, 293.
22. M. D. Romero, G. Ovejero, A. Rodriguez, J. M. Gomez, I. Agueda, *Ind. Eng. Chem. Res.* 2004, 43, 8194.
23. R. L. Guerrero, I. A. Rivero, *ARKIVOC* 2008, 11, 295.
24. V. V. Rekha, M. V. Ramani, A. Ratnamala, V. Rupakalpana, G. V. Subbaraju, C. Satyanarayana, C. Someswara Rao, *Org. Process Res. Dev.* 2009, 13, 769.
25. W.-C. Shieh, S. Dell, O. Repic, *Tetrahedron Lett.* 2002, 43, 5607.
26. F. Rajabi, M. R. Saidi, *Synth. Commun.* 2004, 34, 4179.
27. W.-C. Shieh, S. Dell, O. Repic, *J. Org. Chem.* 2002, 67, 2188.
28. T. N. Glasnov, J. D. Holbrey, C. O. Kappe, K. R. Seddon, T. Yan, *Green Chem.* 2012, 14, 3071.
29. S. R. Kirumakki, N. Nagaraju, K. V. R. Chary, S. Narayanan, *Appl. Catal. A Gen.* 2003, 248, 161.
30. A. Dhakshinamoorthy, A. Sharmila, K. Pitchumani, *Chem. Eur. J.* 2010, 16, 1128.
31. M. Selva, P. Tundo, D. Brunelli, A. Perosa, *Green Chem.* 2007, 9, 463.

10 Synthesis of Ethyl Octyl Ether by Reaction between 1-Octanol and Ethanol over Amberlyst 70

Jordi Guilera, Eliana Ramírez, Montserrat Iborra,
Javier Tejero, and Fidel Cunill

CONTENTS

INTRODUCTION

Linear dialkyl ethers have good properties as fuel components. In addition, they have been used as solvents and also in the formulation and manufacture of dyes, paints, rubbers, resins, and lubricants [1–4]. Typically, dialkyl ethers are produced by using toxic and corrosive chemicals such as alkyl halides or alkyl sulfates, i.e., Williamson ether synthesis. Unfortunately, such reactants form stoichiometric quantities of undesired inorganic salts [5–9]. However, some cleaner technological processes are available using solid acid-base catalysts. Thus, linear dialkyl ethers can be also produced by the dehydration of linear alcohols [10–12] or by the decarboxylation of dialkyl carbonates [13–22]. Alternatively, short-chain dialkyl ethers can also be used as alkylating agents to produce dialkyl ethers of longer chains [18].

Diethyl carbonate has proven to be highly efficient in the ethylation of 1-octanol to form ethyl octyl ether (EOE). We have previously reported yields of 59% over Dowex 50Wx2 using a 2:1 (1-octanol:diethyl carbonate) initial molar ratio at 150°C [22]. Parrot et al. reported yields up to 94% over γ-alumina using an equimolar mixture of reactants at 221°C [19], while Tundo et al. quoted a yield of 99% over basic Al_2O_3

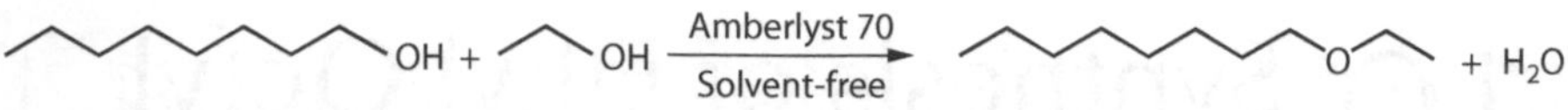

$$\text{octanol-OH} + \text{ethanol-OH} \xrightarrow[\text{Solvent-free}]{\text{Amberlyst 70}} \text{ethyl octyl ether} + H_2O$$

SCHEME 10.1   Target reaction of ethyl octyl ether synthesis from 1-octanol and ethanol.

using a large excess of 1-octanol at 200°C [15]. A drawback of using diethyl carbonate to form ethyl octyl ether is the formation of stoichiometric quantities of CO_2, what increases the environmental impact of the process.

Interestingly, the dehydration reaction between 1-octanol and ethanol also forms ethyl octyl ether by giving water as a by-product. In addition, the current availability of ethanol suggests that its use can be economically advantageous to form ethyl octyl ether. Similar yields (~60%) were obtained from ethanol and 1-octanol over Dowex 50Wx2, with respect to those obtained from diethyl carbonate and 1-octanol in the same experimental conditions [22]. In the present work, the 1-octanol and ethanol reaction to form ethyl octyl ether is evaluated by using stoichiometric amounts of reactants, thus potentially saving energy and materials costs associated with recovery. Besides, the reactor temperature was increased to 190°C by the use of the thermally stable resin Amberlyst 70 (Scheme 10.1).

REACTION PROCEDURE

REPRESENTATIVE PROCEDURE FOR THE SYNTHESIS OF ETHYL OCTYL ETHER BY USING 1-OCTANOL AND ETHANOL OVER AMBERLYST 70

A mixture of 44.0 g of 1-octanol and 15.6 g of ethanol was loaded into a stainless steel autoclave. Then, the reactor was pressurized at 25 bar with N_2 to maintain the liquid phase, heated up to 190°C by an electric furnace, and stirred at 500 rpm. When the mixture reached the working temperature, 1 g of dried Amberlyst 70 was injected into the reactor by shifting with N_2. 0.2 μl liquid samples were analyzed online in a gas–liquid chromatography (GLC). After 48 h of reaction, the reactor was cooled down to room temperature. Ethyl octyl ether was purified by vacuum rectification to 99% (w/w). For more details, see the Notes section.

The reaction scheme of ethyl octyl ether synthesis is shown in Scheme 10.2. Co-etherification of 1-octanol and ethanol yielded ethyl octyl ether and water as products (a). In addition, diethyl ether and di-n-octyl ether were obtained as by-products from the intermolecular dehydration of two ethanol or two 1-octanol molecules, respectively (b, c). Intramolecular dehydration of 1-octanol formed octenes (d). Some amount of ethyl octyl ether could be possibly formed by the reaction of diethyl ether and 1-octanol, and some amount of octenes from the decomposition of ethyl octyl ether or di-n-octyl ether [19].

NOTES

1. Setup and the analysis procedure are described in detail elsewhere [21].
2. 1-Octanol (≥0.995 mass fraction) and ethanol (≥0.998 mass fraction) were purchased from Fluka and Panreac, respectively. Amberlyst 70 was kindly provided by Rohm and Haas France.

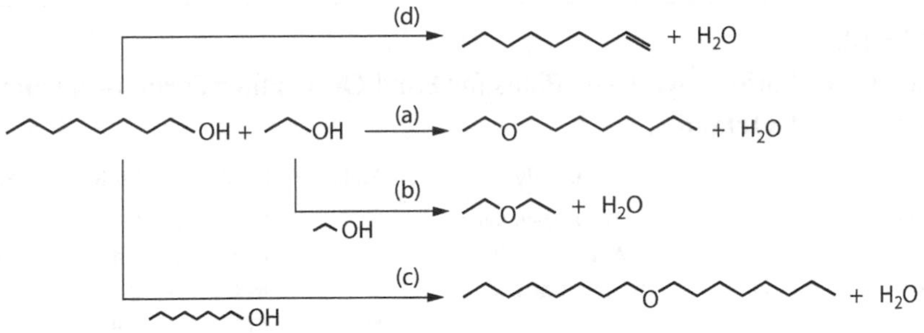

SCHEME 10.2   Reaction scheme of ethyl octyl ether from ethanol and 1-octanol.

3. Sieved Amberlyst 70 ($d_p = 0.49 \pm 0.05$ mm, 95% confidence interval) was used.
4. Catalyst was dried first at 110°C at 1 bar and later at 110°C under vacuum overnight. This procedure assures ≤ 2.25% (w/w) of residual water in Amberlyst 70. The water content was determined by a Karl-Fisher titrator (Orion AF8).
5. Desulfonation of Amberlyst 70 is not significant at 190°C. The catalytic activity of Amberlyst 70 to form ethyl octyl ether is fully recovered as soon as water is removed from the resin [23].
6. A 1 m distillation column (Fisher Scientific) packed with Pall rings was used to purify ethyl octyl ether. Pressure was set at 0.1 bar.

GREEN METRICS ANALYSIS

Calculations for the synthesis of ethyl octyl ether via ethanol and 1-octanol over Amberlyst 70 are shown below.

In Table 10.1, a summary of several synthesis pathways for ethyl octyl ether formation from ethanol and 1-octanol along with their main green metrics parameters is presented for the sake of comparison.

Input (g)	
1-Octanol	44.0
Ethanol	15.6
Amberlyst 70	1
Total	60.6
Output (g)	
Ethyl octyl ether	25.4
Unreacted compounds	13.5
Di-n-octyl ether	10.1
Octenes	4.7
Water	4.1
Diethyl ether	1.8
Amberlyst 70	1
Total	60.6

TABLE 10.1
Summary of Various Synthesis Plans for Ethyl Octyl Ether Formation from Ethanol and 1-Octanol

Plan	Catalyst	% Yield	% AE	E-Factor	PMI
Parrot (2010) [19]	Acidic γ-alumina	94	63.7	0.8	1.8
Guilera (this work)	Amberlyst 70	48	89.8	1.4	2.4
Guilera (2012) EtOH [22]	Dowex 50Wx2	57	89.8	2.5	3.5
Devaney (1953) [5]	None	50	57.3	2.9	3.9
Guilera (2012) DEC[a] [22]	Dowex 50Wx2	59	63.7	3.2	4.2
Casas (2013) [12]	Amberlyst 121	12	89.7	9.6	10.6
Tundo (2004) [15]	Basic alumina	99	63.7	13.1	14.1
Barry (1984) method A [7]	Aliquat 336	86	60.6	27.8	28.8
Barry B (1984) method B [7]	Aliquat 336	80	56.7	30.2	31.2
Urata (1991) [9]	$CO_2(CO)_8$	15	58.5	125.6	126.6
McKillop (1974) [6]	$Hg(ClO_4)_2$	95	66.1	189.4	190.4
Urata (1991) [8]	$CO_2(CO)_8$	1	55.2	12,280	12,281

[a] DEC = diethyl carbonate.

CONCLUSION

Ethyl octyl ether can be successfully produced by using environmentally benign reactants in the temperature range of 150–220°C; that excludes the need to use toxic and corrosive reactants to form linear ethers. The formation of ethyl octyl ether by the dehydration of 1-octanol and ethanol is a simple and solvent-free reaction process that provides the highest atom economy. The use of Amberlyst 70 allows an easy separation and recovery of the catalysts for further use in a batch reactor, or else a long-life performance in a continuous fixed-bed reactor. Moreover, di-n-octyl ether is the main by-product formed (on a weight basis), which is also a highly valuable product. On the other hand, higher mass intensity to ethyl octyl ether can be obtained by the reaction of 1-octanol with diethyl carbonate over γ-alumina; thus, reduced amounts of by-products and unreacted compounds would be necessary to treat. A drawback of the use of diethyl carbonate is the formation of stoichiometric quantities of carbon dioxide.

ACKNOWLEDGMENTS

This work was economically supported by the State Education, Universities, Research & Development Office of Spain (Project CTQ2010-16047).

REFERENCES

1. G. A. Olah, U.S. Patent, U.S. 5520710 A, 1996.
2. R. Patrini, M. Marchionna, U.S. Patent, U.S. 6218583 B1, 2001.
3. K. Seidel, C. Priebe, D. Hollenberg, U.S. Patent, U.S. 5830483, 1998.

4. A. Ansmann, B. Fabry, U.S. Patent, U.S. 6878379 B2, 2005.
5. L. W. Devaney, G. W. Panian, *J. Am. Chem. Soc.* 1953, 75, 4836–4837.
6. A. McKillop, M. E. Ford, *Tetrahedron*, 1974, 30, 2467–2475.
7. J. Barry, G. Bram, G. Decodts, A. Loupy, P. Pigeon, J. Sansoulet. *Tetrahedron* 1984, 40, 2945–2950.
8. H. Urata, H. Maekawa, S. Takahashi, T. Fuchikami, *J. Org. Chem.* 1991, 56, 4320–4322.
9. H. Urata, D. Goto, T. Fuchikami, *Tetrahedron Lett.* 1991, 32, 3091–3094.
10. W. K. Gray, F. R. Smail, M. G. Hitzler, S. K. Ross, M. Poliakoff, *J. Am. Chem. Soc.* 1999, 121, 10711–10718.
11. B. Walsh, J. R. Hyde, P. Licence, M. Poliakoff, *Green Chem.* 2005, 7, 456–463.
12. C. Casas, J. Guilera, E. Ramírez, R. Bringué, M. Iborra, J. Tejero, *Biomass Conv. Bioref.* 2013, 3, 27–37.
13. Y. Ono, *Appl. Catal. A* 1997, 155, 133–166.
14. P. Tundo, M. Selva, *Acc. Chem. Res.* 2002, 35, 706–716.
15. P. Tundo, S. Memoli, D. Hérault, K. Hill. *Green Chem.* 2004, 6, 609–612.
16. P. Tundo, L. Rossi, A. Loris, *J. Org. Chem.* 2005, 70, 2219–2224.
17. M. Selva, A. Perosa, *Green Chem.* 2008, 10, 457–464.
18. P. N. Gooden, R. A. Bourne, A. J. Parrott, H. S. Bevinakatti, D. J. Irvine, M. Poliakoff, *Org. Proc. Res. Dev.* 2010, 14, 411–416.
19. A. J. Parrott, R. A. Bourne, P. N. Gooden, H. S. Bevinakatti, M. Poliakoff, D. J. Irvine, *Org. Proc. Res. Dev.* 2010, 14, 1420–1426.
20. M. Selva, M. Fabris, A. Perosa, *Green Chem.* 2011, 13, 863–872.
21. J. Guilera, R. Bringué, E. Ramírez, M. Iborra, J. Tejero. *Appl. Catal. A* 2012, 413–414, 21–29.
22. J. Guilera, R. Bringué, E. Ramírez, M. Iborra, J. Tejero, *Ind. Eng. Chem. Res.* 2012, 51, 16525–16530.
23. J. Guilera, E. Ramírez, C. Fité, M. Iborra, J. Tejero, *Appl. Catal. A* 2013, 467, 301–309.

11 Reaction: Organocatalytic Asymmetric Tandem Epoxidation–Passerini Reaction

Arlene G. Corrêa, Márcio W. Paixão,
Anna M. Deobald, and Daniel G. Rivera

CONTENTS

INTRODUCTION

Tandem approaches have proven to be a very successful strategy employed in organic synthesis, as they have the potential to rapidly increase molecular complexity by forming several covalent bonds in a one-pot fashion.[1] The main advantages of such tandem protocols are the reduction of manual operations—thus circumventing the isolation of often highly reactive intermediates—and the minimal use of protecting groups. Indeed, both factors help in saving both time and cost, as well as improving the greener character of the whole procedure.[2]

An interesting example is the one-pot, asymmetric multicatalytic Michael addition formal [3+2] reactions between 1,3-dicarbonyls and α,β-unsaturated aldehydes described by Lathrop and Rovis.[3] In this multicatalytic cascade, α-hydroxycyclopentanones containing three stereocenters, including a quaternary stereocenter, could be accessed efficiently by merging aminocatalysis with N-heterocyclic carbene-based catalysis (Scheme 11.1).

In recent years, asymmetric organocatalytic multicomponent reactions have been used in the synthesis of enantiomerically enriched molecules.[4] Krishna and Lopinti have described a diastereoselective three-component Passerini reaction of chiral 2,3-epoxy aldehydes, tosylmethyl isocyanide, and benzoic acid to afford α-acyloxy-β,γ-epoxy-carboxamides in moderate to good yields as well as diastereomeric excess.[5] The chiral epoxyaldehyde partners were synthesized in a previous reaction by using the standard Sharpless asymmetric epoxidation.[6]

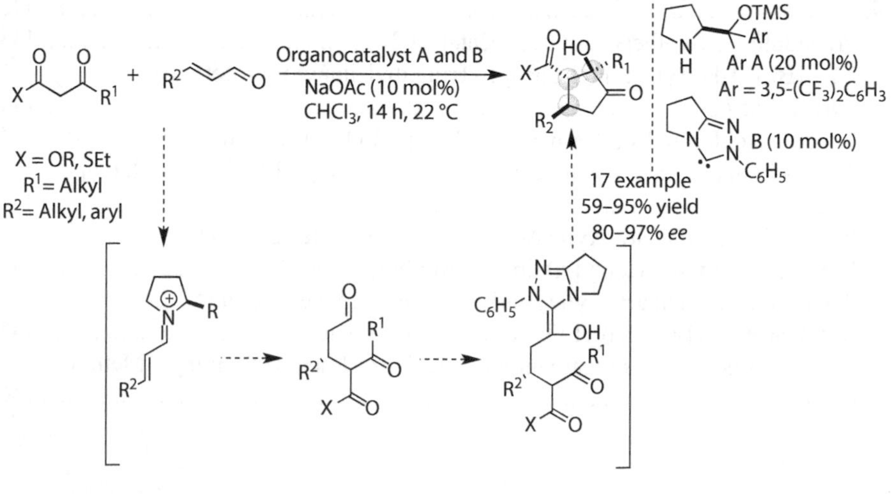

SCHEME 11.1 Asymmetric multicatalytic Michael formal [3+2] reactions.[3]

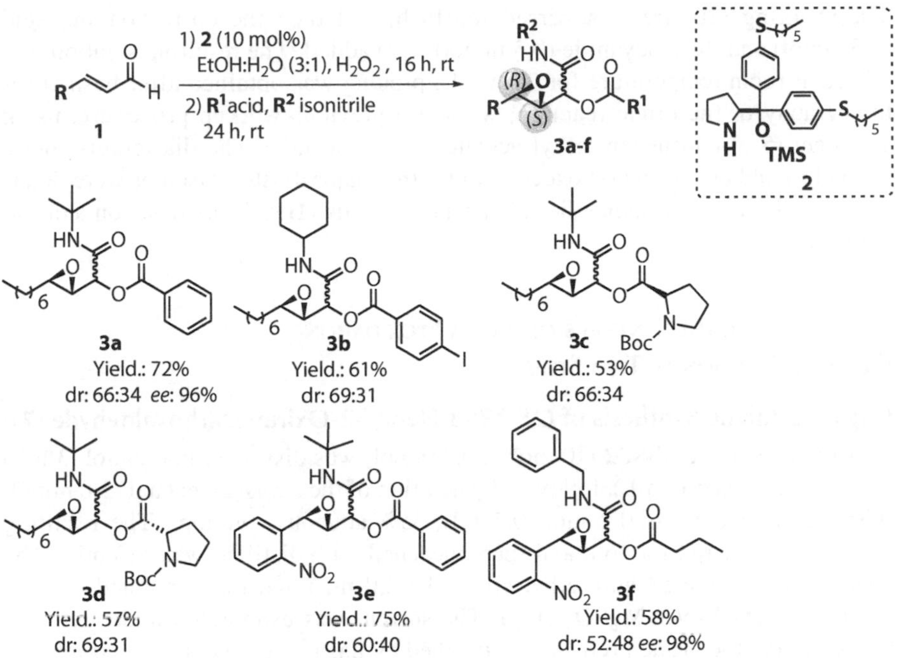

SCHEME 11.2 Organocatalytic asymmetric tandem epoxidation–Passerini reaction.

A very convenient strategy for the synthesis of highly functionalized epoxides was developed by combining an asymmetric organocatalytic epoxidation and the Passerini three-component reaction.[7] This process was carried out employing a diarylprolinolsilyl ether **2**, which showed high organocatalytic activity in an environmentally benign solvent system. A typical protocol is the organocatalytic epoxidation of *trans*-2-decenal (**1**) with 10% mol of catalyst **2** followed by condensation with *tert*-butyl isocyanide and benzoic acid in EtOH/H$_2$O (3:1, v/v) as the solvent system in a concentration of 1 mol L^{-1} of the starting material.

The whole procedure could be performed in one pot without isolation of intermediates, thus affording epoxide **3a** in 72% yield over two steps and a diastereoisomeric ratio (d.r.) of 66:34, with 96% *ee* of the major diastereoisomer. When the process was carried out in two steps, with the isolation of the epoxyaldehyde intermediate, compound **3a** was obtained in 83% global yield; however, it was necessary to employ an additional purification process. In total, six new epoxides were prepared in moderate to good yields using this strategy (Scheme 11.2).

REACTION PROCEDURE

GENERAL PROCEDURE TO ONE-POT EPOXIDATION–PASSERINI REACTION

In a 5 mL vial, the catalyst **2** (10 mol%, 0.03 mmol) was dissolved in ethanol (0.45 mL) and distilled water (0.15 mL), followed by addition of the α,β-unsaturated aldehyde (0.3 mmol) and H$_2$O$_2$ (0.9 mmol, 0.075 mL, 35% aqueous solution).

The resulting mixture was stirred for 16 h, and then the corresponding acid (0.3 mmol) and the isocyanide (0.5 mmol) were added. The resulting solution was stirred at room temperature for 24 h. The product was obtained after flash chromatography of the crude reaction, without a previous workup procedure, using silica gel (5 g) and hexane:ethyl acetate = 4:1 (100 mL). The diastereoisomeric ratio (d.r.) and enantiomeric excess (*ee*) for the major diastereoisomer were determined by high-performance liquid chromatography (HPLC) analysis on a chiral stationary phase.

Procedure for the Synthesis of 3a via Epoxidation– Passerini Reaction in Two Steps

Organocatalytic Synthesis of (2S,3R)-3-Heptyl-2-Oxiranecarboxaldehyde (7)[8]

In a 10 mL vial, catalyst **2** (10 mol%, 0.2 mmol) was dissolved in ethanol (3 mL) and distilled water (1 mL), followed by addition of the *trans*-2-decenal (2.0 mmol, 0.36 mL) and H_2O_2 (6.0 mmol, 0.5 mL, 35% aqueous solution). The resulting mixture was stirred for 16 h and then quenched with distilled water (3 mL). The product was extracted with ethyl ether (3 × 20 mL) and the combined organic phase was dried over $MgSO_4$ (1 g). The solvent was evaporated under reduced pressure, and the crude product was purified by flash column chromatography on silica gel (8 g) and hexane:ethyl acetate = 4:1 (100 mL) with 0.1% of Et_3N. The (2S,3R)-3-heptyl-2-oxiranecarboxaldehyde (**7**) was obtained as a white solid in 90% yield (31 mg), 94:6 d.r., and 90% *ee* (determined on the major diastereoisomer by HPLC analysis on a chiral stationary phase after reduction to the corresponding alcohol).

Multicomponent Approach for the Synthesis of Benzoic Acid *tert*-Butylcarbamoyl-(3-Heptyl-Oxiranyl), Methyl Ester (3a)

In a 5 mL vial, epoxyaldehyde (0.3 mmol) was dissolved in ethanol (0.45 mL) and distilled water (0.15 mL), followed by addition of benzoic acid (0.3 mmol, 0.037 g) and *tert*-butyl isocyanide (0.5 mmol, 0.042 g). The resulting solution was stirred at room temperature for 24 h. Compound **3a** was obtained in 93% yield (105 mg) after flash chromatography of the crude reaction, without a previous workup procedure, using silica gel (5 g) and hexane:ethyl acetate = 4:1 (100 mL).

NOTES

[1]H and ^{13}C nuclear magnetic resonance (NMR) spectra were recorded on a Bruker ARX-400 (400 and 100 MHz, respectively). All NMR spectra were obtained with $CDCl_3$. HPLC chromatograms were obtained on a Shimadzu apparatus, LC-10AT Pump, SPD-10A UV-Vis Detector, SCL-10A System Controller, using a Chiralpak AD-H (4.6 mmØ × 250 mml, particle size 5 μm) or a Chiralcel OD-H or OD (4.6 mmØ × 250 mml, particle size 5 μm). Optical rotations were measured with a Perkin-Elmer Polarimeter, Model 241 at 589 nm and 30°C. Melting point was

obtained on an MQAPF-301 apparatus. High-resolution mass spectra were recorded on a Bruker-AutoFlex Speed, MALDI-TOF/TOF MS ($\lambda = 355$ nm, $f = 500$ Hz, matrix α-cyano-4-hydroxy cinnamic acid (HCCA), calibration standard triphenylphosphate (TPP), polyethylene glycol (PEG) 600). Column chromatography was performed using Merck Silica Gel (230–400 mesh). Thin-layer chromatography (TLC) was performed using Merck Silica Gel GF_{254}, 0.25 mm thickness. For visualization, TLC plates were either placed under ultraviolet light or stained with iodine vapor or acidic vanillin.

1. The following solvents were dried and purified by distillation from the reagents indicated: tetrahydrofuran from sodium with a benzophenone ketyl indicator and dichloromethane from calcium hydride. All other solvents were American Chemical Society (ACS) grade unless otherwise noted. Air- and moisture-sensitive reactions were conducted in flame-dried or oven-dried glassware equipped with tightly fitted rubber septa and under a positive atmosphere of dry argon. Reagents and solvents were handled using standard syringe techniques.
2. The commercial aldehydes were obtained from Sigma-Aldrich.
3. The synthetic route for organocatalyst **2** is outlined in Scheme 11.3.[7] 1-Bromo-4-hexylthiobenzene[9] and N-Boc-L-proline methyl ester (**4**)[10] were prepared as described in the literature.

PROCEDURE FOR SYNTHESIS OF ORGANOCATALYST 2

Step 1: To a round bottom flask under a nitrogen atmosphere were added magnesium strips (20 mmol, 0.48 g), a small crystal of iodine, and dry tetrahydrofuran (THF) (30 mL). The reaction mixture was heated to reflux. A solution of 1-bromo-4-hexylthiobenzene[10] (20 mmol) in dry THF (6 mL) was added dropwise over 30 min. After addition, the reaction mixture was continued to stirring at reflux for additional 2 h and cooled to room temperature. Then a solution of (S)-1-*tert*-butyl-2-methyl

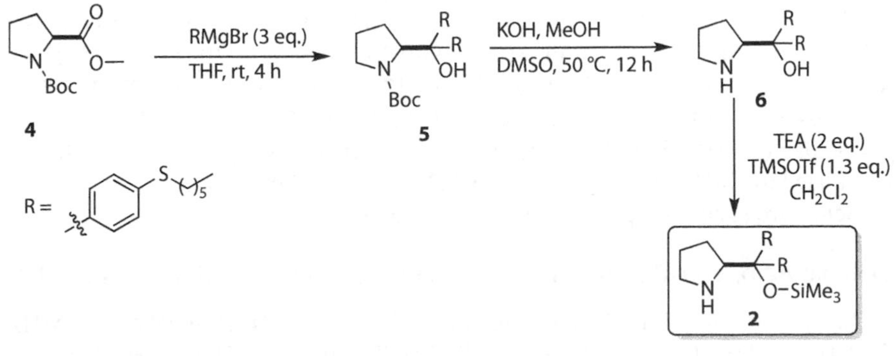

SCHEME 11.3 Synthesis of organocatalyst **2**.

pyrrolidine-1,2-dicarboxylate (**4**) (6.7 mmol, 1.52 g) in dry THF (6 mL) was added dropwise to the Grignard reagent at room temperature over 30 min. The resulting mixture was further stirred for 4 h and then quenched with saturated aqueous solution of NH$_4$Cl (30 mL). The product was extracted with ethyl acetate (3 × 20 mL), and the combined organic phase was dried over MgSO$_4$ (1 g). The solvent was evaporated under reduced pressure, and the crude product was purified by flash column chromatography on silica gel (8 g) and hexane:ethyl acetate = 4:1 (100 mL) to give the product **5** (6.3 mmol, 3.692 g, 95% yield).

Step 2: Compound **5** (4.35 mmol) and KOH (53 mmol, 2.98 g) were dissolved in methanol (6 mL) and dimethyl sulfoxide (DMSO) (27 mL). The reaction mixture was stirred at 50°C for 12 h and was then quenched with water (50 mL). The product was extracted with CH$_2$Cl$_2$ (3 × 30 mL). The organic phase was dried over anhydrous MgSO$_4$ (2 g) and evaporated under reduced pressure, and the crude product was purified by flash column chromatography on silica gel (6 g) and ethyl acetate:methanol = 9:1 (150 mL) to give the product **6** (3.9 mmol, 1.894 g, 90% yield).

Step 3: To a round bottom flask under a nitrogen atmosphere was added the pure aminoalcohol **6** (2.6 mmol) in dry CH$_2$Cl$_2$ (18 mL) and triethylamine (3.4 mmol, 0.47 mL); the reaction mixture was cooled to 0°C. Trimethylsilyltrifluoromethanesulfonate (TMSOTf) (3.4 mmol, 0.61 mL) was added dropwise. The resulting mixture was warmed to room temperature and stirred for 2 h. The reaction mixture was quenched with water (20 mL) and extracted with CH$_2$Cl$_2$ (3 × 30 mL). The organic phase was dried over anhydrous MgSO$_4$ (2 g) and evaporated under reduced pressure. The crude product was purified by column chromatography on silica gel (8 g) using 0.1% triethylamine in hexane:ethyl acetate = 4:1 (100 mL) to give organocatalyst **2** (2.5 mmol, 1.395 g, 96% yield).

2(S)-[BIS-(4-HEXYLSULFANYL-PHENYL)-TRIMETHYLSILANYLOXY-METHYL]-PYRROLIDINE (2)

Global yield: 80%, light yellow oil. ^{1}H NMR (CDCl$_3$, 400 MHz) δ = 7.34 (d, J = 8 Hz, 2H), 7.26–7.19 (m, 6H), 3.99–3.96 (m, 1H), 3.72 (q, J = 7 Hz, 1H), 2.92–2.88 (m, 4H), 2.85–2.80 (m, 1H), 2.77–2.72 (m, 1H), 1.67–1.59 (m, 5H), 1.58–1.54 (m, 2H), 1.45–1.37 (m, 5H), 1.30–1.24 (m, 8H), 0.89–0.86 (m, 6H), −0.10 (s, 9H). ^{13}C NMR (CDCl$_3$, 100 MHz) δ = 143.89, 142.82, 135.70, 128.92, 128.20, 127.86, 127.61, 125.99, 65.31, 64.36, 47.10, 33.43, 31.32, 29.05, 28.48, 27.47, 25.07, 22.49, 13.99, 2.16. [α]$_D$ = −0.025 (c = 0.005 g/mL, CH$_2$Cl$_2$). high-resolution mass spectrometry (HRMS): Calculated to C$_{32}$H$_{51}$NOS$_2$Si [M + H]$^+$ 558.3181; found, 558.3254.

Below are spectroscopic data of epoxides **3a–3f**.

BENZOIC ACID, *TERT*-BUTYLCARBAMOYL-(3-HEPTYL-OXIRANYL)-METHYL ESTER (3A)

Yield: 72%. White solid. Melting point (m.p.) 92–93°C. ^{1}H NMR (CDCl$_3$, 400 MHz) δ = 8.09–8.07 (m, 6H), 7.64–7.59 (m, 3H), 7.51–7.46 (m, 6H), 5.96 (s, 1H), 5.29–5.27 (m, 2H), 4.96 (d, J = 9 Hz, 1H), 3.43 (dd, ^1J = 9 Hz, ^2J = 4 Hz, 1H), 3.29 (dd, ^1J = 5 Hz,

$^2J = 2$ Hz, 1H), 3.21 (dd, $^1J = 5$ Hz, $^2J = 2$ Hz, 1H), 3.08 (td, $^1J = 6$ Hz, $^2J = 2$ Hz, 2H), 3.02 (td, $^1J = 6$ Hz, $^2J = 2$ Hz, 1H), 1.62–1.53 (m, 3H), 1.48–1.41 (m, 3H), 1.39–1.37 (m, 27H), 1.32–1.25 (m, 15H), 0.89–0.85 (m, 9H). ^{13}C NMR (CDCl$_3$, 100 MHz) δ = 166.21, 165.95, 165.66, 164.91, 133.77,133.62, 133.47, 129.99, 129.83, 129.60, 129.07, 128.86, 128.67, 128.57, 128.43, 73.54,73.02, 70.43, 57.83, 57.34, 57.28, 56.61, 56.45, 55.01, 51.75, 51.62, 31.67, 31.59,31.40, 31.35, 29.20, 29.13, 29.10, 28.76, 28.66, 27.57, 26.34, 25.73, 22.59, 14.03. $[\alpha]_D$ = +21.14 (c = 0.007 g/mL, ethyl acetate). HRMS: Calculated to C$_{22}$H$_{33}$NO$_4$ [M + H]$^+$376.2487; found, 376.2481.

4-Iodo-Benzoic Acid, Cyclohexylcarbamoyl-(3-Heptyl-Oxiranyl)-Methyl Ester (3b)

Yield: 56%. White solid. M.p. 122–123°C. ^{1}H NMR (CDCl$_3$, 400 MHz) δ = 7.86–7.83 (m, 6H), 7.79–7.75 (m, 6H), 5.98–5.96 (m, 2H), 5.31 (d, $J = 5$ Hz, 2H), 5.28 (d, $J = 5$ Hz, 1H), 3.84–3.75 (m, 3H), 3.27 (dd, $^1J = 5$ Hz, $^2J = 2$ Hz, 1H), 3.21 (dd, $^1J = 5$ Hz, $^2J = 2$ Hz, 2H), 3.14–3.11 (m, 1H), 3.06 (td, $^1J = 6$ Hz, $^2J = 2$ Hz, 1H), 3.02 (td, $^1J = 6$ Hz, $^2J = 2$ Hz, 1H), 1.94–1.88 (m, 6H), 1.71–1.64 (m, 6H), 1.61–1.55 (m, 9H), 1.47–1.11 (m, 46H), 0.89–0.85 (m, 9H). ^{13}C NMR (CDCl$_3$, 100 MHz) δ = 165.60, 165.32, 164.64, 164.57, 138.06, 137.98, 131.18, 128.49, 128.27, 101.85,101.67, 73.89, 73.19, 57.35, 57.11, 56.70, 56.23, 48.32, 32.86, 32.81, 31.69, 31.35,31.32, 29.21, 29.14, 25.73, 25.68, 25.41, 24.65, 22.60, 14.06. $[\alpha]_D$ = +17.561 (c = 0.008 g/mL, ethyl acetate). HRMS: Calculated to C$_{24}$H$_{34}$INO$_4$ [M + H]$^+$ 528.1610; found, 528.1597.

Pyrrolidine-1,2-Dicarboxylic Acid, 1-tert-Butyl Ester 2S-[tert-Butylcarbamoyl-(3-Heptyl-Oxiranyl)-Methyl] Ester (3c)

Yield: 43%. Colorless oil. ^{1}H NMR (CDCl$_3$, 400 MHz) δ = 6.81 (s, 1H), 6.70 (s, 1H), 5.86 (s, 1H), 5.32 (d, $J = 3$ Hz, 1H), 5.17(d, $J = 3$ Hz, 1H), 5.00 (d, $J = 5$ Hz, 1H), 4.96 (d, $J = 6$ Hz, 1H), 4.78(d, $J = 6$ Hz, 1H), 4.39 (dd, $^1J = 8$ Hz, $^2J = 4$ Hz, 2H), 4.32 (dd, $^1J = 8$Hz, $^2J = 4$ Hz, 1H), 4.26 (dd, $^1J = 8$ Hz, $^2J = 5$ Hz, 2H), 3.59–3.38 (m, 10H), 3.19–3.18 (m, 1H), 3.14–3.12 (m, 3H), 3.09–3.08 (m, 1H), 3.05–3.02 (m, 3H), 2.96–2.93 (m, 2H), 2.34–1.97 (m, 15H), 1.93–1.87 (m, 10H), 1.49 (s, 25H), 1.44–1.35 (m, 80H), 1.31–1.26 (m, 40H), 0.89–0.86 (m, 15H). ^{13}C NMR (CDCl$_3$, 100 MHz) δ = 172.43, 172.10, 171.32, 170.93, 166.04, 165.97, 155.00, 154.44, 80.19, 80.14, 80.11, 74.32, 72.64, 71.87, 59.03, 58.97, 57.44, 57.22, 56.51, 56.12, 55.28, 51.74, 51.63, 50.15, 46.87, 46.72, 46.43, 31.71, 31.67, 31.48, 31.43, 31.34, 30.82, 30.10, 29.96, 29.60, 29.26, 29.20, 29.14, 29.11, 28.62, 28.54, 28.48, 28.33, 25.85, 25.74, 25.70, 24.83, 24.31, 23.65, 22.60, 14.06. $[\alpha]_D$ = –8.636 (c = 0.004 g/mL, ethyl acetate). HRMS: Calculated to C$_{25}$H$_{44}$N$_2$O$_6$ [M + H]$^+$ 469.3277; found, 469.3192.

Pyrrolidine-1,2-Dicarboxylic Acid, 1-tert-Butyl Ester 2R-[tert-Butylcarbamoyl-(3-Heptyl-Oxiranyl)-Methyl] Ester (3d)

Yield: 52%. Colorless oil. ^{1}H NMR (CDCl$_3$, 400 MHz) δ = 6.98 (s, 1H), 6.52 (s, 1H), 5.87 (s, 1H), 5.82 (s, 1H), 5.33 (d, $J = 3$ Hz, 1H), 5.18 (d, $J = 3$ Hz, 1H), 5.16

(d, J = 4 Hz, 1H), 5.04 (d, J = 7 Hz, 1H), 5.00 (d, J = 5 Hz, 1H), 4.96 (d, J = 6 Hz, 1H), 4.78 (d, J = 6 Hz, 1H), 4.38 (dd, ^1J = 8 Hz, ^2J = 4 Hz, 4H), 4.35–4.30 (m, 1H), 4.24 (dd, ^1J = 8 Hz, ^2J = 5 Hz, 2H), 3.58–3.49 (m, 7H), 3.46–3.37 (m, 7H), 3.24–3.23 (m, 4H), 3.20–3.18 (m, 1H), 3.16–3.13 (m, 1H), 3.10–3.06 (m, 4H), 3.03–3.00 (m, 5H), 2.97–2.93 (m, 2H), 2.33–3.15 (m, 7H), 2.12–1.97 (m, 14H), 1.95–1.66 (m, 10H), 1.61–1.53 (m, 8H), 1.50 (s, 14), 1.45–1.42 (m, 56H), 1.40–1.36 (m, 70H), 1.31–1.22 (m, 56H), 0.89–0.86 (m, 21H). ^{13}C NMR (CDCl$_3$, 100 MHz) δ = 172.09, 171.38, 171.13, 166.01, 165.94, 155.10, 154.46, 153.59, 80.22, 80.11, 75.64, 72.09, 70.95, 60.36, 59.07, 59.03, 59.00, 58.97, 58.83, 57.36, 56.93, 56.54, 56.33, 56.18, 55.18, 53.41, 51.77, 51.65, 46.84, 46.69, 46.38, 31.69, 31.48, 31.42, 30.98, 30.25, 29.92, 29.60, 29.23, 29.20, 29.14, 29.11, 28.60, 28.53, 28.44, 28.33, 25.76, 24.73, 24.41, 23.56, 22.60, 21.01, 14.18, 14.04. $[\alpha]_D$ = +11.905 (c = 0.004 g/mL, ethyl acetate). HRMS: Calculated to C$_{25}$H$_{44}$N$_2$O$_6$ [M + H]$^+$ 469.3277; found, 469.3192.

BENZOIC ACID, *TERT*-BUTYLCARBAMOYL-[3-(2-NITRO-PHENYL)-OXIRANYL]-METHYL ESTER (3E)

Yield: 75%. Yellowish solid. M.p. 131–132°C. ^{1}H NMR (CDCl$_3$, 400 MHz) δ = 8.19–8.09 (m, 6H), 7.70–7.59 (m, 6H), 7.53–7.47 (m, 6H), 6.17 (s, 1H), 6.05 (s, 1H), 5.83 (d, J = 2 Hz, 1H), 5.59 (d, J = 4 Hz, 1H), 4.66 (d, J = 2 Hz, 1H), 4.62 (d, J = 2 Hz, 1H), 3.57–3.56 (m, 2H), 1.41–1.40 (m, 18H). ^{13}C NMR (CDCl$_3$, 100 MHz) δ = 165.44, 165.40, 164.87, 164.44, 147.57, 147.48, 134.52, 134.48, 133.93, 133.87, 133.24, 132.91, 129.99, 129.81, 129.74, 129.03, 128.85, 128.81, 128.73, 128.48, 127.15, 127.01, 124.87, 124.84, 72.09, 70.53, 60.48, 56.63, 54.36, 53.28, 51.95, 51.84, 28.68, 28.65. $[\alpha]D$ = +94.44 (c = 0.004 g/mL, ethyl acetate). HRMS: Calculated to C$_{21}$H$_{22}$N$_2$O$_6$ [M + Na]$^+$ 421.13759; found, 421.1374.

BUTYRIC ACID, BENZYLCARBAMOYL-[3-(2-NITRO-PHENYL)-OXIRANYL]-METHYL ESTER (3F)

Yield: 53%. Light yellow solid. M.p. 124–125°C. ^{1}H NMR (CDCl$_3$, 400 MHz) δ = 8.17 (d, J = 8 Hz, 2H), 7.69–7.65 (m, 2H), 7.61–7.58 (m, 2H), 7.53–7.48 (m, 2H), 7.38–7.29 (m, 10H), 6.55 (t, J = 5 Hz, 1H), 6.45 (t, J = 5 Hz, 1H), 5.72 (d, J = 3 Hz, 1H), 5.54 (d, J = 4 Hz, 1H), 4.63–4.45 (m, 6H), 3.50–3.47 (m, 2H), 2.48–2.40 (m, 4H), 1.74–1.66 (m, 4H), 0.99–0.94 (m, 6H). ^{13}C NMR (CDCl$_3$, 100 MHz) δ = 171.82, 171.51, 166.48, 147.55, 147.43, 137.77, 134.48, 134.24, 134.06, 133.02, 132.75, 129.04, 128.89, 128.78, 127.88, 127.71, 127.66, 127.01, 125.05, 124.88, 124.80, 71.79, 71.18, 69.91, 69.55, 60.48, 59.90, 58.43, 57.00, 56.68, 56.54, 54.18, 53.19, 43.48, 35.83, 35.78, 35.31, 31.59, 22.65, 22.47, 18.32, 18.03, 14.12, 13.58, 13.53. $[\alpha]_D$ = +68.831 (c = 0.008 g/mL, ethyl acetate). HRMS: Calculated to C$_{21}$H$_{22}$N$_2$O$_6$ [M + Na]$^+$ 421.1376; found, 421.1385.

GREEN METRICS ANALYSIS

Table 11.1 summarizes the essential material efficiency metrics[11] for the synthesis of **3a** by the conventional two-step and the one-pot strategy, respectively.

TABLE 11.1

Green Metrics Comparison of the Tandem One-Step and Conventional Two-Step Strategies to Synthesize Benzoic Acid, *tert*-Butylcarbamoyl-(3-Heptyl-Oxiranyl)-Methyl Ester (3a)

Compound	Method[a]	% AE[b]	% Yield	E-Kernel	E-Excess	E-Aux	E-Total	PMI[c]
3a	A	95	72	0.5	0.4	942.4	943.3	944.3
7	B, step 1	90	90	0.2	6.1	4220.7	4227.0	4228.0
3a	B, step 2	100	93	0.1	0.2	726.3	726.6	727.6
3a	B, combined step 1 and step 2	95	84	0.2	3.2	2786.3	2789.7	2790.7

[a] Method A = one-pot tandem epoxidation–Passerini reaction; method B = two-step procedure with isolation of epoxide intermediate (2S,3R)-3-heptyl-2-oxiranecarboxaldehyde (7).
[b] AE = atom economy.
[c] PMI = process mass intensity.

TABLE 11.2

Green Metrics Summary for the Synthesis of Organocatalyst 2

Compound	Step	% AE[a]	% Yield	E-Kernel	E-Excess	E-Aux	E-Total	PMI[b]
5	1	63	95	0.7	2.4	53.0	56.1	57.1
6	2	79	90	0.4	2.4	181.9	184.8	185.8
2	3	69	96	0.5	0.2	175.7	176.4	177.4
2	Combined steps 1 to 3	43	82	1.7	5.3	405.4	412.4	413.4

[a] AE = atom economy.
[b] PMI = process mass intensity.

These results show that the tandem one-pot strategy reduces the overall process mass intensity by about threefold when compared with the overall two-step process involving isolation of oxirane (**7**), thus confirming its superior material efficiency performance. The main savings in generating waste materials originate in the auxiliary materials used in the workup and purification phases of the procedure. In the two-step sequence, the first epoxidation step generates six times more waste than the three-component Passerini reaction since no workup materials are used in the second condensation step. Table 11.2 summarizes the materials' green metrics performance for each step in the synthesis of organocatalyst **2**, and overall for the three steps in the linear sequence.

Though the first step involving a Grignard reaction on an ester substrate has the least atom economy, overall it has the lowest process mass intensity (PMI) since it consumes the least mass of auxiliary materials in the workup and purification stages. The second step involving N-Boc deprotection is the most wasteful since it has a high contribution from E-aux to its corresponding E-total. Over the three steps the overall atom economy, yield, and PMI are 43%, 82%, and 413.4, respectively.

REFERENCES

1. Ł. Albrech, H. Jiang, K. A. Jørgensen, *Angew. Chem. Int. Ed.* 2011, 50, 8492–8509.
2. (a) A. Dömling, W. Wang, K. Wang, *Chem. Rev.* 2012, 112, 3083–3135; (b) D. J. Ramon, M. Yus, *Angew. Chem. Int. Ed.* 2005, 44, 1602–1634; (c) A. Dömling, I. Ugi, *Angew. Chem. Int. Ed.* 2000, 39, 3168–3210.
3. S. P. Lathrop, T. Rovis, *J. Am. Chem. Soc.* 2009, 131, 13628–13630.
4. (a) H. Jiang, P. Elsner, K. L. Jensen, A. Falcicchio, V. Marcos, K. A. Jørgensen, *Angew. Chem. Int. Ed.* 2009, 48, 6844–6848; (b) K. C. Nicolaou, D. J. Edmonds, P. G. Bulger, *Angew. Chem. Int. Ed.* 2006, 45, 7134–7186; (c) C. J. Chapman, C. G. Frost, *Synthesis* 2007, 1–21; (d) A. M. Walji, D. W. C. MacMillan, *Synlett* 2007, 1477–1489; (e) S. Lin, L. Deiana, A. Tseggai, A. Córdova, *Eur. J. Org. Chem.* 2012, 398–408.
5. P. R. Krishna, K. Lopinti, *Synlett* 2007, 83–86.
6. P. R. Krishna, L. Krishnarao, V. Kannan, *Tetrahedron Lett.* 2004, 45, 7847–7850.
7. A. M. Deobald, A. G. Correa, D. G. Rivera, M. W. Paixão, *Org. Biomol. Chem.* 2012, 10, 7681–7684.
8. L. Albrecht, H. Jiang, G. Dickmeiss, B. Gschwend, S. G. Hansen, K. A. Jørgensen, *J. Am. Chem. Soc.* 2010, 132, 9188–9196.
9. S. N. Crane, W. C. Black, J. T. Palmer, D. E. Davis, E. Setti, J. Robichaud, J. Paquet, R. M. Oballa, C. I. Bayly, D. J. McKay, J. R. Somoza, N. Chauret, C. Seto, J. Scheigetz, G. Wesolowski, F. Masse, S. Desmarais, M. Ouellet, *J. Med. Chem.* 2006, 49, 1066–1079.
10. X. Ding, M. D. Vera, B. Liang, Y. Zhao, M. S. Leonard, M. M. Joullie, *Bioorg. Med. Chem. Lett.* 2001, 11, 231–234.
11. (a) C. Jiménez-González, C. S. Ponder, Q. B. Broxterman, J. B. Manley, *Org. Proc. Res. Dev.* 2011, 15, 912–917; (b) J. Andraos, *Org. Proc. Res. Dev.* 2009, 13, 161–185.

12 Reaction: Green Synthesis of Chalcone and Coumarin Derivatives via a Suzuki Cross-Coupling Reaction

Lucas C. C. Vieira, Márcio W. Paixão, and Arlene G. Corrêa

CONTENTS

INTRODUCTION

Chalcones and coumarins are important naturally occurring plant constituents and display a wide range of pharmacological and biological activities.[1,2] From a synthetic perspective, chalcone and coumarin derivatives are gaining more prominence due to their pivotal role in organic synthesis.[3] However, only a few methods have been already reported for the derivatization of this class of compounds under Suzuki cross-coupling conditions.[4]

A substantial number of reports have appeared in the literature for the synthesis of 3-, 4-, 6-, and 7-substituted coumarins by Suzuki coupling,[5] but none for the 8-substituted coumarins. Moreover, these methodologies usually employ conventional organic solvents[6] and relatively unstable palladium catalysts,[7] and they require

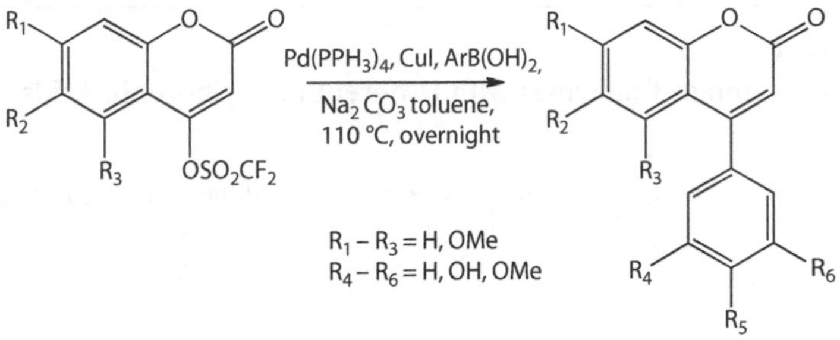

SCHEME 12.1 Cross-coupling reaction of aryl boronic acids and 4-trifluoromethylsulfonyloxycoumarins.[9]

harsh reaction conditions and long reaction times.[8] An example is the synthesis of 4-arylcoumarins by a cross-coupling reaction of aryl boronic acids and 4-trifluoromethylsulfonyloxycoumarins in the presence of $Pd(PPh_3)_4$ and copper(I) iodide as a co-catalyst under reflux in dry toluene overnight (Scheme 12.1). The resulting neoflavones were isolated in good to excellent yields (59% to 99%).[9]

On account of these aspects, an eco-friendly, efficient, and versatile approach for the Suzuki coupling reaction of chalcones (Table 12.1) and coumarins (Table 12.2) with aryl boronic acids with low catalyst loading using nontoxic polyethylene glycol (PEG) as a solvent was developed (Scheme 12.2). Furthermore, the reaction conditions employed include an economically inexpensive and stable catalyst, $Pd(OAc)_2$, microwave irradiation as a source of energy, and short reaction times, thus satisfying the principles of green chemistry.[10]

REACTION PROCEDURE

GENERAL PROCEDURE FOR THE SUZUKI CROSS-COUPLING REACTION

Chalcone **1a–e** or coumarin **4** (0.151 mmol), aryl boronic acid (0.226 mmol), KF (26.2 mg, 0.453 mmol), $Pd(OAc)_2$ (3.4 mg, 10 mol%), PEG-400 (3 mL, 2.65 g), and ethanol (1 mL) were placed in a glass tube and irradiated for 30–45 min in a CEM Discovery® focused microwave oven at 110°C. Then, the crude was filtered through a sintered glass plate funnel containing a pad of silica gel (5 g) with ethyl acetate (25 mL). The organic phase was washed with water (2 × 10 mL) and brine (2 × 10 mL), dried over $MgSO_4$ (200 mg), and concentrated under vacuum. Chalcones **2a–s** (Table 12.1) were purified by recrystallization from EtOH (3 mL) and coumarins **5a–f** (Table 12.2) by column chromatography in silica gel (10 g) using hexane:ethyl acetate 9:1 as eluent (150 mL).

CONVENTIONAL PROCEDURE[11]

To a 10 mL round-bottom flask was added coumarin **4** (51.5 mg, 0.143 mmol), $Pd(OAc)_2$ (1.6 mg, 0.005 mmol), PPh_3 (3.7 mg, 0.01 mmol), and boronic acid (0.165 mmol). The flask was evacuated and backfilled with argon four times. The mixture was dissolved in dimethyl formamide (DMF) (2 mL) and stirred at room temperature for 15 min.

TABLE 12.1
Suzuki Reaction of Chalcones with Different Phenylboronic Acids[a]

Entry	Chalcone	Product	R_1	R_2	Yield (%)[c]
1	1a	2a	H	H	92
2	1a	2b	H	OCH_3	95
3	1a	2c	NO_2	H	94
4	1a	2d	OCH_3	H	90
5	1a	2e	F	OCH_3	87
6	1a	2f	CH_3	F	94
7	1a	2g	$O-CH_2-O$	91	
8	1b	2h[b]	H	H	91
9	1b	2i[b]	H	OCH_3	95
10	1b	2j[b]	NO_2	H	84
11	1b	2k[b]	F	OCH_3	90
12	1b	2l[b]	CH_3	F	91
13	1c	2m	H	OCH_3	91
14	1c	2n	$O-CH_2-O$	90	
15	1d	2o	H	OCH_3	81
16	1d	2p	NO_2	H	79
17	1e	2q	H	OCH_3	53
18	1e	2r	OCH_3	H	50
19	1e	2s	NO_2	H	52

[a] Reaction conditions: chalcone (1.0 eq.), boronic acid (1.5 eq.), base (3 eq.), ethanol (1 mL), and PEG-400 (3 mL).
[b] Reaction performed in 45 min.
[c] Isolated yields.

Potassium carbonate (59.2 mg, 0.429 mmol) was dissolved in degassed water (2 mL) and then added to the reaction mixture, which was stirred at room temperature for 4 h. Afterwards, the mixture was filtered through celite (5 g) with ethyl acetate (20 mL) and the organic layer was concentrated. The residue was purified using flash column chromatography on silica gel (9 g) and 30% ethyl acetate in dichloromethane (150 mL), affording coumarins **5a** and **5c–e** (Table 12.3).

NOTES

Unless otherwise noted, all commercially available reagents were purchased from Aldrich Chemical Co. and used without purification. [1]H and ^{13}C nuclear magnetic resonance (NMR) spectra were recorded on a Bruker ARX-400 (400 and 100 MHz,

TABLE 12.2
Suzuki Reaction of Coumarin 4 with Different Phenylboronic Acids[a]

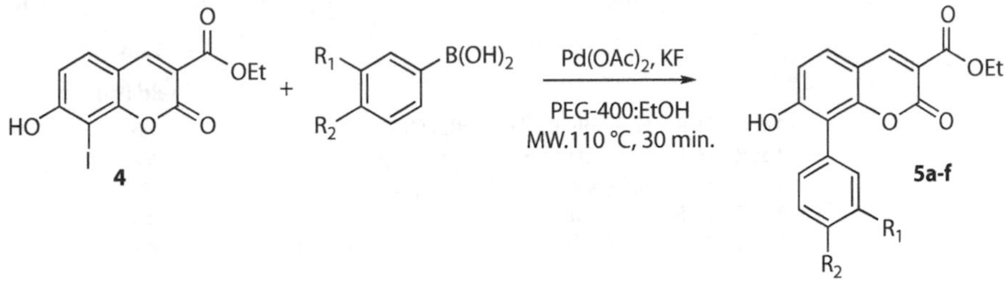

Coumarin	R_1	R_2	Yield (%)[b]
5a	H	H	95
5b	H	OMe	85
5c	NO_2	H	72
5d	OMe	H	87
5e	H	CH_2OH	78
5f	Cl	H	80

[a] Reaction conditions: coumarin (1.0 eq.), boronic acid (1.5 eq.), KF (3 eq.), ethanol (1 mL), and PEG-400 (3 mL).

[b] Isolated yields.

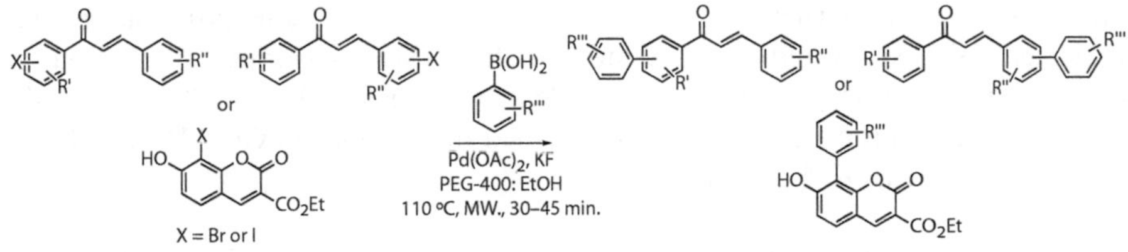

SCHEME 12.2 Suzuki coupling reaction of chalcones and coumarins.

TABLE 12.3
Reaction Conditions for Conventional Procedure

Coumarin	Boronic Acid (mg)	Yield (%), mg
5a	Phenyl (20.1)	50, 22.7
5c	3-Nitro-phenyl (27.5)	57, 28.8
5d	3-Methoxy-phenyl (25.0)	66, 32.0
5e	4-Hydroxymethyl-phenyl (25.0)	70, 34.0

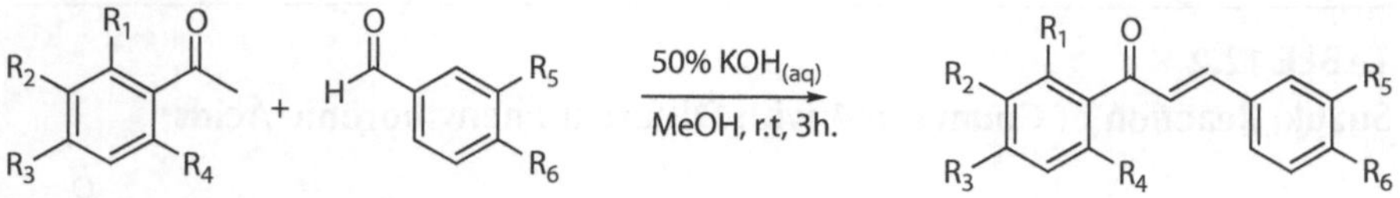

1a-e

Yield (%)

1a: R_1=H, R_2=H, R_3=Br, R_4=H, R_5=R_6=OCH$_2$O   95
1b: R_1=H, R_2=R_3=OCH$_2$O, R_4=H, R_5=H, R_6=Br   92
1c: R_1=H, R_2=H, R_3=Br, R_4=H, R_5=H, R_6=OCH$_3$ 90
1d: R_1=H, R_2=H, R_3=Br, R_4=H, R_5=H, R_6=H   83
1e: R_1=OH, R_2=I, R_3=OCH$_3$, R_4=OCH$_3$, R_5=H, R_6=OCH$_3$ 62

SCHEME 12.3   Synthesis of chalcones **1a–e**.

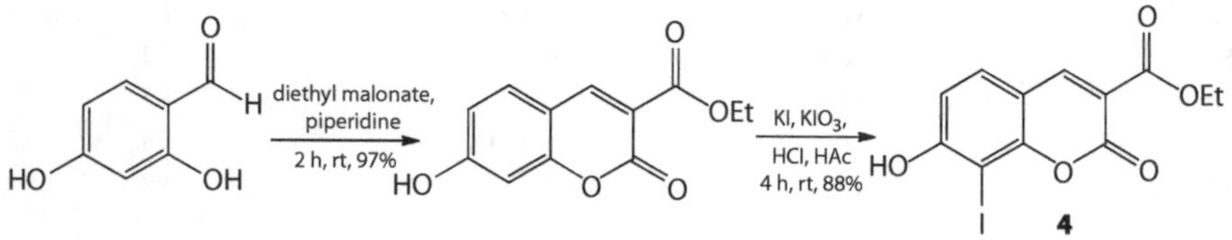

SCHEME 12.4   Synthesis of coumarin **4**.

respectively). Mass spectra were recorded on a Shimadzu GCMS-QP5000. Elemental analyses were performed on a Fisons EA 1108 CHNS-O. Analytical thin-layer chromatography was performed on a 0.25 mm film of silica gel containing fluorescent indicator UV$_{254}$ supported on an aluminum sheet (Sigma-Aldrich). Flash column chromatography was performed using silica gel (Kieselgel 60, 230–400 mesh, E. Merck). Gas chromatography was performed in a Shimadzu GC-17A with N$_2$ as carrier and using a DB-5 column.

1. Chalcones **1a–e** (Scheme 12.3) were synthesized via Claisen-Smith condensation reaction with 3,4-substituted benzaldehyde and an aromatic acetophenone, aqueous KOH (50% w/v) in methanol, at room temperature for 3 h.[12] After completion of the reaction, the mixture was filtered to collect the precipitates and purification by recrystallization afforded the pure chalcones **1a–e** in 62% to 95% yield.
2. Coumarin **4** was prepared via Knoevenagel condensation of 2-hydroxy-benzaldehyde[13] with diethyl malonate followed by iodination reaction[14] (Scheme 12.4).

ETHYL 7-HYDROXY-2-OXO-2H-CHROMENE-3-CARBOXYLATE[13]

To a solution of 2,4-dihydroxybenzaldehyde (511 mg, 3.70 mmol) in diethyl malonate (1.26 g, 7.90 mmol), piperidine (10 drops, about 1 mL) was added, and the resulting solution was stirred for 2 h at room temperature. The resulting solution was then acidified with a 10% aqueous solution of HCl (5 mL). The precipitate was filtrated and washed with cold water (10 mL). The crude product was purified by flash chromatography using silica gel (50 g) and dichloromethane:ethyl acetate = 7:3 (200 mL) as eluent, furnishing the desired 7-hydroxyl-coumarin (0.735 g) in 85% yield.

^{1}H NMR (400 MHz, DMSO-$d6$) δ: 1.28 (t, J = 7.1 Hz, 3 H), 4.24 (q, J = 7.1 Hz, 2 H), 6.71 (d, J = 1.8 Hz, 1 H), 6.83 (dd, J = 1.8, 8.3 Hz, 1 H), 7.74 (d, J = 8.3 Hz, 1 H), 8.66 (s, 1 H), 11.07 (s, OH). ^{13}C NMR (100 MHz, DMSO-$d6$) δ: 14.1, 60.8, 101.7, 110.4, 112.0, 113.9, 132.1, 149.4, 156.3, 157.0, 162.9, 164.0.

ETHYL 7-HYDROXY-8-IODO-2-OXO-2H-CHROMENE-3-CARBOXYLATE (4)[14]

A solution of **3** (232 mg, 0.991 mmol), potassium iodide (110 mg, 0.660 mmol), and potassium iodate (71 mg, 0.330 mmol) was prepared in acetic acid (6 mL). The mixture was treated at room temperature with a solution of hydrochloric acid (36 mg, 0.991 mmol) in water (5 mL) over 45 min and stirred for an additional 4 h. The precipitate was filtrated and washed with saturated sodium bisulfite (10 mL) and cold water (10 mL). The product was purified by recrystallization from methanol (5 mL) furnishing **4** (0.310 g) in 88% yield.

^{1}H NMR (400 MHz, DMSO-$d6$) δ: 1.30 (t, J = 7.1 Hz, 3 H), 4.27 (q, J = 7.2 Hz, 2 H), 6.95 (d, J = 8.6 Hz, 1 H), 7.75 (d, J = 8.8 Hz, 1 H), 8.65 (s, 1 H), 11.89 (s, 1 H). ^{13}C NMR (100 MHz, DMSO-$d6$) δ: 14.0, 60.8, 73.7, 111.1, 112.5, 131.6, 149.2, 156.1, 156.6, 162.6, 163.6. Elemental analysis: calculated: C, 40.02; H, 2.52; found: C, 40.05; H, 2.50.

Below are spectroscopic data of chalcone and coumarin derivatives:

(E)-3-(BENZO[D][1,3]DIOXOL-5-YL)-1-(BIPHENYL-4-YL)PROP-2-EN-1-ONE (2A)[15]

Yellow solid (45.5 mg, 92%). ^{1}H NMR (400 MHz, CDCl$_3$) δ: 6.03 (s, 2 H), 6.85 (d, J = 8.1 Hz, 1 H), 7.14 (dd, J = 1.6, 8.3 Hz, 1 H), 7.19 (d, J = 1.8 Hz, 1 H), 7.49–7.40 (m, 5 H), 7.65 (d, J = 7.5 Hz, 2 H), 7.72 (d, J = 8.3 Hz, 2 H), 7.77 (d, J = 15.6 Hz, 1 H), 8.09 (d, J = 8.1 Hz, 2 H). ^{13}C NMR (100 MHz, CDCl$_3$) δ: 101.7, 106.7, 108.7, 120.1, 125.3, 127.3, 127.3, 128.2, 129.0, 129.1, 129.4, 137.1, 140.0, 144.7, 145.4, 148.5, 150.0, 189.8. Elemental analysis: calculated: C, 80.47; H, 4.91; found: C, 80.43; H, 4.95. MS (relative intensity %) m/z: 328 (100), 152 (46), 122 (44), 89 (28).

(E)-3-(BENZO[D][1,3]DIOXOL-5-YL)-1-(4'-METHOXYBIPHENYL-4-YL)PROP-2-EN-1-ONE (2B)

Yellow solid (51.3 mg, 95%). ^{1}H NMR (400 MHz, CDCl$_3$) δ: 3.86 (s, 3 H), 6.03 (s, 2 H), 6.85 (d, J = 8.1 Hz, 1 H), 7.01 (d, J = 8.9 Hz, 2 H), 7.14 (dd, J = 8.1, 1.3 Hz, 1 H), 7.19 (d, J = 1.3 Hz, 1 H), 7.41 (d, J = 15.6 Hz, 1 H), 7.60 (d, J = 8.6 Hz, 2 H), 7.68 (d, J = 8.3 Hz, 2 H), 7.76 (d, J = 15.6 Hz, 1 H), 8.07 (d, J = 8.6 Hz, 2 H). ^{13}C NMR (100 MHz, CDCl$_3$) δ: 55.4, 101.7, 106.7, 108.7, 114.4, 120.1, 125.2, 126.7, 128.4, 129.1, 129.5, 132.4, 136.5, 144.5, 145.0, 148.4, 149.9, 159.9, 189.7. Elemental analysis: calculated: C, 70.08; H, 5.06; found: C, 70.11; H, 5.09. MS (relative intensity %) m/z: 358 (100), 236 (24), 150 (22), 89 (23).

(E)-3-(Benzo[d][1,3]Dioxol-5-yl)-1-(3'-Nitrobiphenyl-4-yl)Prop-2-en-1-One (2c)

Yellow solid (52.9 mg, 94%). ^{1}H NMR (400 MHz, CDCl$_3$) δ: 6.04 (s, 2 H), 6.86 (d, J = 8.1 Hz, 1 H), 7.16 (dd, J = 8.1, 1.6 Hz, 1 H), 7.20 (d, J = 1.6 Hz, 1 H), 7.40 (d, J = 15.6 Hz, 1 H), 7.66 (t, J = 7.9 Hz, 1 H), 7.76 (d, J = 8.6 Hz, 2 H), 7.79 (d, J = 15.3 Hz, 1 H), 7.99–7.96 (m, 1 H), 8.13 (d, J = 8.3 Hz, 2 H), 8.26 (ddd, J = 8.3, 2.2, 1.1 Hz, 1 H), 8.51 (t, J = 1.9 Hz, 1 H). ^{13}C NMR (100 MHz,CDCl$_3$) δ: 101.7, 106.7, 108.8, 119.8, 122.2, 122.9, 125.5, 127.4, 129.3, 129.3, 130.0, 133.2, 138.3, 141.7, 142.6, 143.0, 145.2, 148.5, 148.8, 189.6. Elemental analysis: calculated: C, 70.77; H, 4.05; N, 3.75; found: C, 70.80; H, 4.08; N, 3.72. MS (relative intensity %) m/z: 373 (100), 145 (41), 122 (65), 89 (46).

(E)-3-(Benzo[d][1,3]Dioxol-5-yl)-1-(3'-Methoxybiphenyl-4-yl)Prop-2-en-1-One (2d)[16]

Yellow solid (48.6 mg, 90%). ^{1}H NMR (400 MHz, CDCl$_3$) δ: 3.88 (s, 3 H), 6.03 (s, 2 H), 6.85 (d, J = 8.1 Hz, 1 H), 6.95 (dd, J = 8.3, 1.9 Hz, 1 H), 7.19–7.13 (m, 3 H), 7.23 (d, J = 7.8 Hz, 1 H), 7.43–7.37 (m, 2 H), 7.71 (d, J = 8.3 Hz, 2 H), 7.77 (d, J = 15.3 Hz, 1 H), 8.08 (d, J = 8.3 Hz, 2 H). ^{13}C NMR (100 MHz, CDCl$_3$) δ: 55.4, 101.7, 106.7, 108.7, 113.1, 113.5, 119.8, 120.0, 125.3, 127.3, 129.0, 129.4, 130.0, 137.2, 141.5, 144.7, 145.3, 148.4, 150.0, 160.1, 189.8. Elemental analysis: calculated: C, 77.08; H, 5.06; found: C, 77.10; H, 5.03. MS (relative intensity %) m/z: 358 (100), 236 (17), 122 (27), 89 (25).

(E)-3-(Benzo[d][1,3]Dioxol-5-yl)-1-(3'-Fluoro-4'-Methoxybiphenyl-4-yl)Prop-2-en-1-One (2e)

Yellow solid (49.4 mg, 87%). ^{1}H NMR (400 MHz, CDCl$_3$) δ: 3.94 (s, 3 H), 6.03 (s, 2 H), 6.85 (d, J = 8.1 Hz, 1 H), 7.07–7.03 (m, 1 H), 7.14 (dd, J = 8.1, 1.6 Hz, 1 H), 7.19 (d, J = 1.6 Hz, 1 H), 7.42–7.37 (m, 3 H), 7.65 (d, J = 8.3 Hz, 2 H), 7.76 (d, J = 15.6 Hz, 1 H), 8.06 (d, J = 8.3 Hz, 2 H). ^{13}C NMR (100 MHz, CDCl$_3$) δ: 56.4, 101.7, 106.7, 108.7, 113.8, 114.8, 115.0, 120.0, 123.0, 125.3, 126.7, 129.2, 129.4, 137.0, 143.8, 144.7, 147.9 (d, J = 10.5 Hz), 148.5, 150.0, 153.9, 189.7. Elemental analysis: calculated: C, 73.40; H, 4.55; found: C, 73.43; H, 4.59. MS (relative intensity %) m/z: 376 (100), 207 (26), 145 (29), 89 (27).

(E)-3-(Benzo[d][1,3]Dioxol-5-yl)-1-(4'-Fluoro-3'-Methylbiphenyl-4-yl)Prop-2-en-1-One (2f)

Yellow solid (51.1 mg, 94%). ^{1}H NMR (400 MHz, CDCl$_3$) δ: 2.36 (s, 3 H), 6.03 (s, 2 H), 6.85 (d, J = 8.1 Hz, 1 H), 7.19–7.07 (m, 3 H), 7.46–7.38 (m, 3 H), 7.66 (d, J = 8.1 Hz, 1 H), 7.76 (d, J = 15.6 Hz, 1 H), 8.07 (d, J = 8.1 Hz, 2 H). ^{13}C NMR (100 MHz, CDCl$_3$) δ: 14.7, 101.6, 106.6, 108.7, 115.5 (d, J = 22.5 Hz), 119.9, 125.2, 125.3 (d, J = 18.3 Hz), 126.1 (d, J = 7.5 Hz), 127.0, 129.0, 129.4, 130.3 (d, J = 6.5 Hz), 135.8, 136.9, 144.5, 144.6, 148.4, 149.9, 161.5 (d, J = 246.9 Hz), 189.6. Elemental analysis: calculated: C, 76.65;

H, 4.75; found: C, 76.66; H, 4.72. MS (relative intensity %) m/z: 360 (100), 165 (32), 122 (51), 89 (36).

(E)-3-(Benzo[d][1,3]Dioxol-5-yl)-1-(4-(Benzo[d][1,3]Dioxol-5-yl)Phenyl) Prop-2-en-1-One (2g)

Yellow solid (51.2 mg, 91%). ^{1}H NMR (400 MHz, CDCl$_3$) δ: 6.03 (s, 2 H), 6.04 (s, 2 H), 6.85 (d, J = 8.1 Hz, 1 H), 6.91 (dd, J = 7.5, 0.8 Hz, 1 H), 7.13–7.12 (m, 2 H), 7.15–7.14 (m, 1 H), 7.19 (d, J = 1.6 Hz, 1 H), 7.41 (d, J = 15.6 Hz, 1 H), 7.64 (d, J = 8.3 Hz, 2 H), 7.76 (d, J = 15.6 Hz, 1 H), 8.06 (d, J = 8.3 Hz, 2 H). ^{13}C NMR (100 MHz, CDCl$_3$) δ: 101.4, 101.7, 106.7, 107.7, 108.7, 108.8, 120.0, 121.1, 125.3, 126.9, 128.9, 129.1, 129.5, 134.3, 136.8, 144.6, 145.1, 147.9, 148.4, 148.5, 189.7. Elemental analysis: calculated: C, 74.19; H, 4.33; found: C, 74.22; H, 4.37. MS (relative intensity %) m/z: 372 (100), 250 (29), 139 (42), 89 (27).

(E)-1-(Benzo[d][1,3]Dioxol-5-yl)-3-(Biphenyl-4-yl)Prop-2-en-1-One (2h)

Yellow solid (45.8 mg, 91%). ^{1}H NMR (400 MHz, CDCl$_3$) δ: 6.08 (s, 2 H), 6.91 (d, J = 8.1 Hz, 1 H), 7.40–7.38 (m, 1 H), 7.51–7.45 (m, 3 H), 7.55 (d, J = 1.2 Hz, 1 H), 7.68–7.62 (m, 5 H), 7.72 (d, J = 8.2 Hz, 2 H), 7.84 (d, J = 15.8 Hz, 1 H). ^{13}C NMR (100 MHz, CDCl$_3$) δ: 101.9, 107.8, 108.5, 121.5, 124.7, 127.0, 127.2, 127.6, 127.9, 128.9, 133.0, 140.1, 143.2, 143.8, 148.3, 151.7, 188.2. Elemental analysis: calculated: C, 80.47; H, 4.91; found: C, 80.49; H, 4.93. MS (relative intensity %) m/z: 328 (100), 251 (27), 178 (34), 121 (36).

(E)-1-(Benzo[d][1,3]Dioxol-5-yl)-3-(4'-Methoxybiphenyl-4-yl)Prop-2-en-1-One (2i)

Yellow solid (51.3 mg, 95%). ^{1}H NMR (400 MHz, CDCl$_3$) δ: 3.86 (s, 3 H), 6.07 (s, 2 H), 6.90 (d, J = 8.3 Hz, 1 H), 7.01–6.98 (m, 2 H), 7.51 (d, J = 15.6 Hz, 1 H), 7.55 (d, J = 1.5 Hz, 1 H), 7.62–7.56 (m, 4 H), 7.70–7.66 (m, 3 H), 7.83 (d, J = 15.6 Hz, 1 H). ^{13}C NMR (100 MHz, CDCl$_3$) δ: 55.4, 101.9, 107.9, 108.4, 114.4, 118.0, 121.2, 124.6, 127.0, 128.1, 128.9, 132.6, 133.3, 142.8, 143.9, 148.3, 151.7, 159.6, 188.2. Elemental analysis: calculated: C, 77.08; H, 5.06; found: C, 77.07; H, 5.09. MS (relative intensity %) m/z: 358 (100), 184 (26), 165 (34), 149 (26).

(E)-1-(Benzo[d][1,3]Dioxol-5-yl)-3-(3'-Nitrobiphenyl-4-yl)Prop-2-en-1-One (2j)

Yellow solid (47.3 mg, 84%). ^{1}H NMR (400 MHz, CDCl$_3$) δ: 6.08 (s, 2H), 6.92 (d, J = 8.0 Hz, 1H), 7.58–7.54 (m, 2 H), 7.70–7.62 (m, 4 H), 7.76 (d, J = 8.3 Hz, 2 H), 7.84 (d, J = 15.6 Hz, 1 H), 7.96–7.94 (m, 1 H), 8.25–8.22 (m, 1 H), 8.49 (m, 1 H). ^{13}C NMR (100 MHz, CDCl$_3$) δ: 101.9, 107.9, 108.4, 121.8, 122.4, 122.5, 124.8, 127.6, 129.1, 129.9, 132.9, 135.3, 140.3, 141.8, 143.1, 148.4, 148.8, 150.2, 151.8, 188.0. Elemental analysis: calculated: C, 70.77; H, 4.05; N, 3.75; found: C, 70.80; H, 4.07; N, 3.79. MS (relative intensity %) m/z: 373 (100), 251 (38), 149 (78), 121 (54).

(*E*)-1-(Benzo[d][1,3]Dioxol-5-yl)-3-(3'-Fluoro-4'-Methoxybiphenyl-4-yl)Prop-2-en-1-One (2k)

Yellow solid (51.1 mg, 90%). ¹H NMR (400 MHz, CDCl₃) δ: 3.94 (s, 3 H), 6.07 (s, 2 H), 6.90 (d, *J* = 8.3 Hz, 1 H), 7.06–7.02 (m, 1 H), 7.39–7.34 (m, 2 H), 7.59–7.49 (m, 4 H), 7.70–7.65 (m, 3 H), 7.82 (d, *J* = 15.5 Hz, 1 H). ¹³C NMR (100 MHz, CDCl₃) δ: 56.3, 101.9, 107.9, 108.4, 113.7, 114.6, 114.8, 121.5, 122.6, 124.7, 127.0, 129.0, 133.0, 133.2, 133.3, 133.9, 141.6, 143.7, 148.3, 151.3, 151.7, 153.8, 188.2. Elemental analysis: calculated: C, 73.40; H, 4.55; found: C, 73.38; H, 4.52. MS (relative intensity %) *m/z*: 376 (100), 174 (28), 149 (38), 121 (26).

(*E*)-1-(Benzo[d][1,3]Dioxol-5-yl)-3-(4'-Fluoro-3'-Methylbiphenyl-4-yl)Prop-2-en-1-One (2l)

Yellow solid (49.4 mg, 91%). ¹H NMR (400 MHz, CDCl₃) δ: 2.35 (s, 3 H), 6.06 (s, 2 H), 6.90 (d, *J* = 8.3 Hz, 1 H), 7.10–7.06 (m, 1 H), 7.43–7.38 (m, 2 H), 7.59–7.49 (m, 4 H), 7.69–7.65 (m, 3 H), 7.82 (d, *J* = 15.5 Hz, 1 H). ¹³C NMR (100 MHz, CDCl₃) δ: 14.7, 101.9, 107.9, 108.4, 115.3, 115.5, 121.5, 124.6, 125.2, 125.9, 126.0, 127.4, 128.9, 130.1, 130.2, 133.0, 133.8, 136.0, 142.4, 143.7, 148.3, 151.7, 160.1, 188.2. Elemental analysis: calculated: C, 76.65; H, 4.75; found: C, 76.69; H, 4.73. MS (relative intensity %) *m/z*: 360 (100), 196 (25), 174 (27), 149 (34).

(*E*)-1-(4'-Methoxybiphenyl-4-yl)-3-(4-Methoxyphenyl)Prop-2-en-1-One (2m)

Yellow solid (49.3 mg, 91%). ¹H NMR (400 MHz, CDCl₃) δ: 3.85 (s, 3 H), 3.86 (s, 3 H), 6.94 (d, *J* = 8.8 Hz, 2 H), 7.00 (d, *J* = 8.8 Hz, 2 H), 7.46 (d, *J* = 15.5 Hz, 1 H), 7.63–7.59 (m, 4 H), 7.67 (d, *J* = 8.3 Hz, 2 H), 7.81 (d, *J* = 15.5 Hz, 1 H), 8.07 (d, *J* = 8.3 Hz, 2 H). ¹³C NMR (100 MHz, CDCl₃) δ: 55.4, 55.5,114.4, 119.7, 126.7, 127.5, 127.7, 128.4, 129.1, 130.3, 132.4, 136.6, 144.5, 145.0, 159.9, 161.7, 189.9. Elemental analysis: calculated: C, 80.21; H, 5.85; found: C, 80.19; H, 5.88. MS (relative intensity %) *m/z*: 344 (100), 329 (22), 236 (29), 161 (29).

(*E*)-1-(4-(Benzo[d][1,3]Dioxol-5-yl)Phenyl)-3-(4-Methoxyphenyl)Prop-2-en-1-One (2n)

Yellow solid (50.8 mg, 90%). ¹H NMR (400 MHz, CDCl₃) δ: 3.86 (s, 3 H), 6.02 (s, 2 H), 6.91 (dd, *J* = 1.0, 7.2 Hz, 1 H), 6.94 (d, *J* = 8.5 Hz, 2 H), 7.15–7.12 (m, 2 H), 7.45 (d, *J* = 15.8 Hz, 1 H), 7.75–7.70 (m, 4 H), 7.81 (d, *J* = 15.8 Hz, 1 H), 8.06 (d, *J* = 8.5 Hz, 2 H). ¹³C NMR (100 MHz, CDCl₃) δ: 55.4, 98.3, 101.3, 107.6, 108.7, 114.4, 119.7, 121.0, 126.8, 127.7, 129.0, 130.2, 134.2, 136.8, 144.5, 145.0, 148.3, 161.7, 189.8. Elemental analysis: calculated: C, 77.08; H, 5.06; found: C, 77.05; H, 5.09. MS (relative intensity %) *m/z*: 358 (100), 250 (31), 161 (30), 139 (43).

(*E*)-1-(4'-Methoxybiphenyl-4-yl)-3-Phenylprop-2-en-1-One (2o)

Yellow solid (44.3 mg, 81%). ¹H NMR (400 MHz, CDCl₃) δ: 3.86 (s, 3 H), 7.01 (d, *J* = 8.8 Hz, 2 H), 7.44–7.41 (m, 3 H), 7.58 (d, *J* = 15.8 Hz, 1 H), 7.60

(d, J = 8.8 Hz, 2 H), 7.70–7.65 (m, 4 H), 7.84 (d, J = 15.8 Hz, 1 H), 8.09 (d, J = 8.5 Hz, 2 H).^{13}C NMR (100 MHz, CDCl$_3$) δ: 55.4, 114.5, 122.1, 126.7, 127.8, 128.4, 128.5, 129.0, 129.2, 130.5, 132.4, 136.3, 144.6, 145.2, 159.9, 189.9. Elemental analysis: calculated: C, 84.05; H, 5.77; found: C, 84.08; H, 5.79. MS (relative intensity %) m/z: 314 (100), 211 (24), 103 (31), 77 (20).

(E)-1-(3'-Nitrobiphenyl-4-yl)-3-Phenylprop-2-en-1-One (2p)

Yellow solid (45.2 mg, 79%). ^{1}H NMR (400 MHz, CDCl$_3$) δ: 7.45–7.43 (m, 3 H), 7.57 (d, J = 15.6 Hz, 1 H), 7.68–7.64 (m, 3 H), 7.77 (d, J = 8.6 Hz, 2 H), 7.86 (d, J = 15.6 Hz, 1 H), 7.98 (dt, J = 7.7, 1.2 Hz, 1 H), 8.15 (d, J = 8.3 Hz, 2 H), 8.26 (ddd, J = 8.3, 2.2, 0.8 Hz, 1 H), 8.51 (t, J = 1.9 Hz, 1 H). ^{13}C NMR (100 MHz, CDCl$_3$) δ: 121.7, 122.1, 122.9, 127.4, 128.5, 129.0, 129.4, 130.0, 130.7, 133.1, 134.7, 138.0, 141.6, 142.7, 145.3, 148.8, 189.7. Elemental analysis: calculated: C, 76.58; H, 4.59; N, 4.25; found: C, 76.53; H, 5.62; N, 4.26. MS (relative intensity %) m/z: 329 (100), 131 (45), 103 (64), 77 (38).

(E)-1-(2-Hydroxy-4,4',6-Trimethoxybiphenyl-3-yl)-3-(4-Methoxyphenyl) Prop-2-en-1-One (2q)

Orange solid (25.2 mg, 53%). ^{1}H NMR (400 MHz, CDCl$_3$) δ: 3.83 (s, 3 H), 3.84 (s, 3 H), 4.02 (s, 3 H), 6.10 (s, 1 H), 6.93 (d, J = 8.9 Hz, 2 H), 6.97 (d, J = 8.9 Hz, 2 H), 7.30 (d, J = 8.9 Hz, 2 H), 7.57 (d, J = 8.9 Hz, 2 H), 7.80 (sl, 2 H), 14.24 (s, 1 H). ^{13}C NMR (100 MHz, CDCl$_3$) δ: 55.2, 55.4, 55.7, 55.9, 86.6, 106.5, 111.3, 113.5, 114.4, 125.1, 125.4, 128.3, 130.1, 132.1, 142.4, 158.4, 161.3, 162.1, 162.9, 164.1, 193.1. Elemental analysis: calculated: C, 71.41; H, 5.75; found: C, 71.43; H, 5.72.

(E)-1-(2-Hydroxy-3',4,6-Trimethoxybiphenyl-3-yl)-3-(4-Methoxyphenyl) Prop-2-en-1-One (2r)

Orange solid (23.8 mg, 50%). ^{1}H NMR (400 MHz, CDCl$_3$) δ: 3.82 (s, 3 H), 3.83 (s, 3 H), 3.85 (s, 3 H), 4.02 (s, 3 H), 6.10 (s, 1 H), 6.87 (ddd, J = 8.2, 2.6, 0.8 Hz, 1 H), 6.96–6.91 (m, 3 H), 7.33 (t, J = 7.9 Hz, 1 H), 7.57 (d, J = 8.9 Hz, 2 H), 7.80 (s, 2 H), 14.22 (s, 1 H).^{13}C NMR (100 MHz, CDCl$_3$) δ: 55.2, 55.4, 55.8, 55.9, 86.7, 106.5, 111.6, 112.6,114.4, 116.8, 123.5, 125.4, 128.3, 128.8, 130.1, 134.5, 142.5, 159.2, 161.4, 162.4, 162.8, 164.0, 193.1. Elemental analysis: calculated: C, 71.41; H, 5.75; found: C, 71.39; H, 5.77.

(E)-1-(2-Hydroxy-4,6-Dimethoxy-3'-Nitrobiphenyl-3-yl)-3-(4-Methoxyphenyl)Prop-2-en-1-One (2s)

Orange solid (25.7 mg, 52%). ^{1}H NMR (400 MHz, CDCl$_3$) δ: 3.86 (sl, 6 H), 4.05 (s, 3 H), 6.12 (s, 1 H), 6.94 (d, J = 8.9 Hz, 2 H), 7.54 (d, J = 7.8 Hz, 1 H), 7.58 (d, J = 8.6 Hz, 2 H), 7.73 (ddd, J = 7.8, 1.3, 1.1 Hz, 1 H), 7.81 (d, J = 1.3 Hz, 2 H), 8.16 (ddd, J = 8.3, 2.4, 1.1 Hz, 1 H), 8.28–8.27 (m, 1 H), 14.49 (s, 1 H). ^{13}C NMR (100 MHz, CDCl$_3$) δ: 55.4, 55.8, 56.0, 86.5, 106.4, 109.1, 110.5, 114.4, 121.8, 125.0, 126.5, 128.2,

128.5, 130.2, 130.3, 134.9, 137.7, 143.1, 148.0, 161.5, 162.5, 193.1. Elemental analysis: calculated: C, 66.20; H, 4.86; N, 3.22; found: C, 66.21; H, 4.89; N, 3.25.

Ethyl 7-Hydroxy-2-Oxo-8-Phenyl-2H-Chromene-3-Carboxylate (5a)

Yellow solid (40.9 mg, 95%). ^{1}H NMR (200 MHz, DMSO-$d6$) δ: 1.28 (t, J = 7.2 Hz, 3 H), 4.25 (q, J = 7.1 Hz, 2 H), 7.03 (d, J = 8.5 Hz, 1 H), 7.46–7.36 (m, 5 H), 7.76 (d, J = 8.7 Hz, 1 H), 8.71 (s, 1 H). ^{13}C NMR (50 MHz, DMSO-$d6$) δ: 14.2, 60.9, 110.7, 111.9, 113.8, 115.7, 127.5, 128.0, 130.7, 130.8, 131.5, 149.7, 154.0, 156.5, 161.4, 163.0. Elemental analysis: calculated: C, 69.67; H, 4.55; found: C, 69.70; H, 4.50.

Ethyl 7-Hydroxy-8-(4-Methoxyphenyl)-2-Oxo-2H-Chromene-3-Carboxylate (5b)

Yellow solid (40.1 mg, 85%). ^{1}H NMR (400 MHz, DMSO-$d6$) δ: 1.27 (t, J = 7.1 Hz, 3 H), 3.80 (s, 3 H), 4.24 (q, J = 7.0 Hz, 2 H), 7.00 (d, J = 8.9 Hz, 3 H), 7.30 (d, J = 8.9 Hz, 2 H), 7.71 (d, J = 8.9 Hz, 1 H), 8.68 (s, 1 H). ^{13}C NMR (100 MHz, DMSO-$d6$) δ: 14.1, 39.5, 55.1, 60.8, 110.7, 111.9, 113.4, 113.6, 115.3, 123.2, 130.4, 131.8, 149.7, 154.1, 156.4, 158.6, 161.3, 162.9. Elemental analysis: calculated: C, 67.05; H, 4.74; found: C, 67.06; H, 4.72.

Ethyl 7-Hydroxy-8-(3-Nitrophenyl)-2-Oxo-2H-Chromene-3-Carboxylate (5c)

Yellow solid (35.5 mg, 72%). ^{1}H NMR (400 MHz, DMSO-$d6$) δ: 1.29 (t, J = 7.1 Hz, 3 H), 4.26 (q, J = 7.1 Hz, 2 H), 7.06 (d, J = 8.6 Hz, 1 H), 7.78 (t, J = 7.9 Hz, 1 H), 7.83 (d, J = 8.6 Hz, 1 H), 7.92 (d, J = 7.8 Hz, 1 H), 8.30–8.25 (m, 2 H), 8.73 (s, 1 H), 11.32 (s, 1 H). ^{13}C NMR (100 MHz, DMSO-$d6$) δ: 14.1, 60.8, 110.7, 112.3, 113.1, 113.6, 122.4, 125.4, 129.4, 131.6, 133.1, 137.6, 147.6, 149.5, 153.9, 156.0, 160.8, 162.8. Elemental analysis: calculated: C, 60.85; H, 3.69; N, 3.94; found: C, 60.81; H, 3.65; N, 3.98.

Ethyl 7-Hydroxy-8-(3-Methoxyphenyl)-2-Oxo-2H-Chromene-3-Carboxylate (5d)

Yellow solid (41.0 mg, 87%). ^{1}H NMR (200 MHz, DMSO-$d6$) δ: 1.27 (t, J = 7.1 Hz, 3 H), 3.76 (s, 3 H), 4.24 (q, J = 7.0 Hz, 2 H), 6.95–6.91 (m, 3 H), 7.01 (d, J = 8.5 Hz, 1 H), 7.35 (t, J = 7.8 Hz, 1 H), 7.74 (d, J = 8.5 Hz, 1 H), 8.68 (s, 1 H), 10.93 (s, OH). ^{13}C NMR (50 MHz, DMSO-$d6$) δ: 14.1, 55.0, 60.7, 110.6, 112.0, 112.8, 113.6, 115.5, 116.4, 122.9, 128.8, 130.7, 132.6, 149.5, 153.9, 156.3, 158.8, 161.1, 162.8. Elemental analysis: calculated: C, 67.05; H, 4.74; found: C, 67.10; H, 4.71.

Ethyl 7-Hydroxy-8-(4-(Hydroxymethyl)Phenyl)-2-Oxo-2H-Chromene-3-Carboxylate (5e)

Yellow solid (36.8 mg, 78%). ^{1}H NMR (400 MHz, DMSO-$d6$) δ: 1.27 (t, J = 6.8 Hz, 3 H), 4.24 (q, J = 6.6 Hz, 2 H), 4.54 (s, 2 H), 7.01 (d, J = 8.3 Hz, 1 H),

7.32 (d, J = 7.7 Hz, 2 H), 7.38 (d, J = 7.7 Hz, 2 H), 7.74 (d, J = 8.6 Hz, 1 H), 8.69 (s, 1 H). ^{13}C NMR (100 MHz, DMSO-$d6$) δ: 14.1, 60.7, 62.8, 110.7, 111.9, 113.6, 115.6, 126.0, 129.6, 130.3, 130.6, 141.7, 149.7, 154.0, 156.3, 161.2, 162.9. Elemental analysis: calculated: C, 67.05; H, 4.74; found: C, 67.03; H, 4.71.

ETHYL 8-(3-CHLOROPHENYL)-7-HYDROXY-2-OXO-2H-CHROMENE-3-CARBOXYLATE (5F)

Yellow solid (38.2 mg, 80%). ^{1}H NMR (400 MHz, DMSO-$d6$) δ: 1.27 (t, J = 7.1 Hz, 3 H), 4.24 (q, J = 7.0 Hz, 2 H), 7.01 (d, J = 8.6 Hz, 1 H), 7.35 (d, J = 7.3 Hz, 1 H), 7.50–7.42 (m, 3 H), 7.76 (d, J = 8.6 Hz, 1 H), 8.69 (s, 1 H). ^{13}C NMR (100 MHz, DMSO-$d6$) δ: 14.1, 60.8, 110.7, 112.1, 113.7, 114.1, 127.5, 129.5, 129.8, 130.4, 131.3, 132.5, 133.6, 149.6, 153.9, 156.3, 161.1, 162.9. Elemental analysis: calculated: C, 62.71; H, 3.80; found: C, 62.68; H, 3.74.

GREEN METRICS ANALYSIS

Table 12.4 summarizes the material green metrics[16] performances for the synthesis of coumarins **5a**, **5c**, **5d**, and **5e** via the conventional and microwave procedures.

In each case, the microwave procedure produces a process mass intensity that is significantly lower than the conventional procedure. The main reductions in waste generation are in the contributions from auxiliary material consumption and excess reagents. In addition, reaction yields under microwave irradiation are also higher than conventional heating. Atom economies in each case again favor the microwave irradiation method since no potassium carbonate base is employed, thus reducing the waste contribution from reaction by-products. Typically the conventional procedure required a base-to-substrate molar ratio of 4 to 1.

TABLE 12.4
Material Green Metrics Comparison of the Synthesis Performance of Conventional and Microwave Procedures for Chromenes 5

Compound	Method[a]	% AE[b]	% Yield	E-Kernel	E-Excess	E-Aux	E-Total	PMI[c]
5a	A	49	51	3.0	89.8	9462.7	9555.6	9556.6
5a	B	64	95	0.6	0.2	4209.5	4210.3	4211.3
5c	A	52	57	2.4	70.9	7458.4	7531.7	7532.7
5c	B	67	72	1.1	0.3	4840.4	4841.7	4842.7
5d	A	51	66	2.0	63.8	6712.6	6778.3	6779.3
5d	B	66	87	0.7	0.3	4200.0	4201.0	4202.0
5e	A	51	70	1.8	60.0	6317.7	6379.6	6380.6
5e	B	66	78	0.9	0.3	4683.3	4684.3	4685.3

[a] Method A = conventional procedure, method B = microwave procedure.
[b] AE = atom economy.
[c] PMI = process mass intensity.

REFERENCES

1. N. K. Sahu, S. S. Balbhadra, J. Choudhary, D. V. Kohli, *Curr. Med. Chem.* 2012, 19, 209–225.
2. C. Kontogiorgis, A. Detsi, D. Hadjipavlou-Litina, *Exp. Op. Ther. Pat.* 2012, 22, 437–454.
3. A. Yu. Fedorov, A. V. Nyuchev, I. P. Beletskaya, *Chem. Heterocycl. Comp.* 2012, 48, 166–178.
4. (a) C. J. Bennett, S. T. Caldwell, D. B. McPhail, P. C. Morrice, G. G. Duthieb, R. C. Hartleya, *Bioorg. Med. Chem.* 2004, 12, 2079–2098. (b) H. Che, H. Lim, H. P. Kim, H. Park, *Eur. J. Med. Chem.* 2011, 46, 4657–4660. (c) K. Dahlén, E. A. A. Wallén, M. Grøtli, K. Luthman, *J. Org. Chem.* 2006, 71, 6863–6871. (d) L. Larsen, D. H. Yoon, R. T. Weavers, *Synth. Commun.* 2009, 39, 2935–2948. (e) R. Bernini, S. Cacchi, I. De Salvea, G. Fabrizi, *Tetrahedron Lett.* 2007, 48, 4973–4976.
5. (a) H. Sakai, T. Hirano, S. Mori, S. Fujii, H. Masuno, M. Kinoshita, H. Kagechika, A. Tanatani, *J. Med. Chem.* 2011, 54, 7055–7065. (b) M. Schiedel, C. Briehn, P. Baüerle, *J. Organomet. Chem.* 2002, 653, 200–208. (c) N. S. Sitnikov, A. S. Shavyrin, G. K. Fukin, I. P. Beletskaya, S. Combes, A. Y. Fedorova, *Russ. Chem. B* 2010, 59, 626–631. (d) M. J. R. P. Queiroz, A. S. Abreu, R. C. Calhelha, M. S. D. Carvalho, P. M. T. Ferreira, *Tetrahedron* 2008, 64, 5139–5146. (e) S. Hesse, G. Kirsch, *Tetrahedron Lett.* 2002, 43, 1213–1215. (f) W. Gao, Y. Luo, Q. Ding, Y. Peng, J. Wua, *Tetrahedron Lett.* 2010, 51, 136–138. (g) M. J. Matos, S. Vazquez-Rodriguez, F. Borges, L. Santana, E. Uriarte, *Tetrahedron Lett.* 2011, 52, 1225–1227.
6. (a) W. Mao, T. Wang, H. Zeng, Z. Wang, J. Chen, J. Shen, *Bioorg. Med. Chem. Lett.* 2009, 19, 4570–4573. (b) K. Li, Y. Zeng, B. Neuenswander, J. A. Tunge, *J. Org. Chem.* 2005, 70, 6515–6518. (c) A. V. Lipeeva, E. E. Shul'ts, M. M. Shakirov, G. A. Tolstikov, *Russ. J. Org. Chem.* 2011, 47, 1404–1409.
7. (a) O. G. Ganina, E. Daras, V. Bourgarel-Rey, V. Peyrot, A. N. Andresyuk, J. Finet, A. Y. Fedorov, I. P. Beletskaya, S. Combes, *Bioorg. Med. Chem.* 2008, 16, 8806–8812. (b) J. Wu, L. A. Zhang, H. Xia, *Tetrahedron Lett.* 2006, 47, 1525–1528.
8. (a) S. Starcevic, P. Brozic, S. Turk, J. Cesar, T. L. Rizner, S. Gobec, *J. Med. Chem.* 2011, 54, 248–261. (b) C. Bailly, C. Bal, P. Barbier, S. Combes, J. Finet, M. Hildebrand, V. Peyrot, N. Wattez, *J. Med. Chem.* 2003, 46, 5437–5444. (c) Q. Zhu, J. Wu, R. Fathi, Z. Yang, *Org. Lett.* 2002, 4, 3333–3336.
9. J.-T. Pierson, A. Dumètre, S. Hutter, F. Delmas, J.-P. Finet, N. Azas, S. Combes, M. Laget, *Eur. J. Med. Chem.* 2010, 45, 864–869.
10. L. C. C. Vieira, M. W. Paixão, A. G. Corrêa, *Tetrahedron Lett.* 2012, 53, 2715–2718.
11. S. A. Moteki, J. M. Takacs, *Angew. Chem. Int. Ed.* 2008, 47, 894–897.
12. V. Cechinel-Filho, Z. R. Vaz, L. Zunino, J. B. Calixto, R. A. Yunes, *Eur. J. Med. Chem.* 1996, 31, 833–839.
12. F. Bigi, L. Chesini, R. Maggi, G. Sartori, *J. Org. Chem.* 1999, 64, 1033–1035.
13. (a) S. Adimurthy, G. Ramachandraiah, P. K. Ghosh, A. V. Bedekar, *Tetrahedron Lett.* 2003, 44, 5099–5101. (b) S. Mistry, S. Desai, S. S. M. Rao, A. Shah, *Indian J. Heterocyclic Chem.* 2004, 13, 301–306.
14. N. P. Buu-Hoi, N. D. Xuong, *Bull. Soc. Chim. Fr.* 1958, 758–761.
15. N. P. Buu-Hoi, M. Sy, *Bull. Soc. Chim. Fr.* 1958, 219–220.
16. (a) C. Jiménez-González, C. S. Ponder, Q. B. Broxterman, J. B. Manley, *Org. Proc. Res. Dev.* 2011, 15, 912–917; (b) J. Andraos, *Org. Proc. Res. Dev.* 2009, 13, 161–185.

13 Enzymatic-Synthesized Hydroxy-L-Prolines as Asymmetric Catalysts for Green Synthesis

Lo'ay A. Al-Momani

CONTENTS

INTRODUCTION

Proline and proline derivatives are well known as organocatalysts.[1] They have been used over a wide range of organic synthesis reactions to obtain enantiomerically or diastereomerically highly enriched asymmetric products.[2–4] The hydroxyprolines and their derivatives are analogues of proline,[5,6] which is the only secondary proteinogenic amino acid. Its rigidity could control the stereochemistry through the formed imine/enamine intermediate.[7–11] L-Proline and its four derivatives—*cis*-3-L-hydroxyproline, *trans*-3-L-hydroxyproline, *cis*-4-L-hydroxyproline, and *trans*-4-L-hydroxyproline—are available in very pure form and are commercially inexpensive. Hydroxyprolines are prepared from L-proline via enzymatic hydroxylation reactions. These hydroxylations were carried out by Hüttel and Klein,[12,13] and were performed on gram scales. Modification and derivatization[14–18] of L-proline—expecting to improve the efficiency of its catalytic behavior—have been the target of many working groups. Amide and peptide derivatives of proline

163

were found as efficient and selective catalysts for some asymmetric organic reactions such as aldol addition.[19–25]

The hydroxyprolines **2–5** and some of their derivatives like **6** are supposed to be very valuable L-proline derivatives. The *trans*-4-isomer and some of its derivatives have been known for a long time and tested as catalysts for aldol addition and other reactions.[9,26,27] Very few studies are reported about the respective *cis*-isomers. Several studies were conducted on the Zn-proline complex as an asymmetric catalyst.[28–30] The hydroxyprolines **2–5** and the *tert*-butyl ether **6** are the targets of this study. In this work, the results of all hydroxyprolines and their derivatives will be compared with L-proline (**1**) under the same conditions. Four asymmetric reactions will be studied: Aldol addition reaction, Mannich reaction, Michael addition reaction, and Knoevenagel reaction. Here, the enzymatic[12] or microbial[31–33] access toward all regio- and stereoisomeric hydroxyprolines is combined with a thorough investigation of the influence on stereoselectivity in asymmetric catalysis.

EXPERIMENTAL SECTION

MATERIALS AND TECHNIQUES

All reagents used were of analytical grade. Solvents were dried by standard methods if necessary. Thin-layer chromatography (TLC) was carried out on aluminum sheets precoated with silica gel 60F254 (Merck). Detection was accomplished by UV light (λ = 254 nm). Preparative column chromatography was carried out on silica gel 60 (Merck, 40–63 μm). ^{1}H nuclear magnetic resonance (NMR) spectra were recorded on an AMX400 (Bruker BioSpin, Freiburg, Germany). $CDCl_3$ (δ = 7.26 ppm), hydrogen deuterium oxide (HDO) (δ = 4.81 ppm), and dimethyl sulfoxide (DMSO) (δ = 2.50 ppm) are used as internal standards. ^{13}C NMR spectra were calibrated with $^{13}CDCl_3$ (δ = 77.00 ppm) and DMSO (δ = 39.43 ppm) as internal standards. *ee* was determined by chiral phase high-performance liquid chromatography (HPLC) (Chiralpak AS-H, Daicel). In the aldol reaction, the eluent is n-hexane:isopropanol (70:30), UV 254 nm, flow rate 0.7 mL/min. R-isomer, $t_R = 13.6$ min; S-isomer, $t_R = 17.3$ min. 25°C. In the Michael addition reaction, the eluent is n-hexane:isopropanol (85:15), UV 254 nm, flow rate 0.7 mL/min. Isomer 1, $t_R = 12.1$ min; isomer 2, $t_R = 14.5$ min. 25°C. In the Mannich reaction, the eluent is n-hexane:isopropanol (90:10), UV 254 nm, flow rate 1.0 mL/min. Isomer 1, $t_R = 30.7$ min; isomer 2, $t_R = 33.5$ min. 25°C.

GENERAL PROCEDURES

Aldol Reaction

Product (9). p-Nitrobenzaldehyde (**7**) (151 mg, 1.0 mmol) was dissolved in 2 mL acetone (**8**). A solution of 23 mg (0.2 mmol) of **1** in 10 mL DMSO was added and the mixture stirred overnight at room temperature. The reaction mixture was diluted with ethyl acetate (50 mL) and washed twice with water (2 × 20 mL). The organic layer (product **9**) was dried over anhydrous Na_2SO_4 (10 g). According to the type of catalyst (**1–6**), 157–209 mg (75% to 100%) of product **9** was isolated (see Table 13.1). ^{1}H NMR (400 MHz, 298 K, ppm, in $CDCl_3$): $\delta = 8.21$ (d, $J = 8.8$ Hz, 2H, H-arom), 7.54 (d, $J = 8.6$ Hz. 2H, H-arom), 5.26 (dt, $J = 3.7$ Hz, $J = 7.6$ Hz, 1H, CH-O), 3.58 (d, $J = 3.3$ Hz, 1H, CO-C*H*H), 2.85 (dd, $J = 3.4$ Hz, $J = 6.1$ Hz, 1H, CO-CH*H*), 2.22 (s, 3H, CO-CH$_3$), 1.59 (bs, 1H, OH). ^{13}C NMR (101 MHz, 298 K, ppm, in $CDCl_3$): $\delta = 208.47$ (C = O), 149.94 (C-arom), 147.26 (C-arom), 126.37 (2 × C-arom), 123.72 (2 × C-arom), 68.86 (C-O), 51.46 (CO-CH_2), 30.68 (CO-CH_3).

TABLE 13.1

Summary of the Results of the Aldol Addition Reaction

		Aldol Addition Reaction		
Entry	Catalyst	Time of Reaction (h)	Conversion (%)	ee (%)/Isomer
1	1	24	100	65/(R)-**9**
2	2	18	91	74/(R)-**9**
3	3	20	75	39/(R)-**9**
4	4	20	86	55/(R)-**9**
5	5	18	100	57/(R)-**9**
6	6	18	89	69/(R)-**9**

Michael Addition

10   11   12

Product (12). Compound **10** (117 mg, 0.8 mmol) was added to 2 equivalents of dimethyl malonate (**11**) (211 mg, 1.6 mmol, 183 µl) and mixed. Two equivalents of piperidine (1.6 mmol, 136 mg, 158 µl) was added to the mixture and stirred at room temperature for 10 min. 5.0 mg (0.04 mmol) of catalyst **1** was added. The reaction mixture was stirred to the completion and diluted with ethyl acetate (10 mL). It was washed twice with water. (2 × 5 mL). The organic layer (product **12**) was dried over anhydrous Na_2SO_4 (2 g). According to the type of catalysts (**1–6**), 186–222 mg (84% to 100%) of product **12** was isolated (see Table 13.2). The reaction was repeated with the same mole ratio using catalysts **2–6**. ^1H NMR (400 MHz, 298 K, ppm, in $CDCl_3$): δ = 7.22–7.09 (m, 5H, H-arom), 3.89 (ddd, J = 5.6 Hz, J = 8.4 Hz, J = 9.6 Hz, 1H, CH_2-CH), 3.65 (s, 3H, O-CH_3), 3.63 (s, 3H, O-CH_3), 3.48 (m, 1H, CO-CH-CO), 3.29 (m, 1H, CO-CHH), 2.87 (dd, J = 5.3 Hz, J = 6.9 Hz, 1H, CO-CHH), 1.94 (s, 3H, CO-CH_3). ^{13}C NMR (101 MHz, 298 K, ppm, in $CDCl_3$): δ = 205.58 (C = O), 168.24 (C = O), 167.96 (C = O), 167.72 (C-arom), 163.75 (C-arom), 140.10 (C-arom), 128.22 (C-arom), 127.68 (C-arom), 126.95 (C-arom), 56.80 (O-CH_3), 52.04 (O-CH_3), 46.81(CO-CH-CO), 40.13(CO-CH_2), 29.99 (CH-CH_2), 24.04 (CO-CH_3).

TABLE 13.2
Summary of the Results of Michael Addition Reaction

| | | Michael Addition Reaction | | |
Entry	Catalyst	Time of Reaction (day)	Conversion (%)	ee (%)
1	1	2	100	25
2	2	2	93	<5
3	3	2	100	27
4	4	2	92	<5
5	5	3	84	<5
6	6	2	100	19

Mannich Product

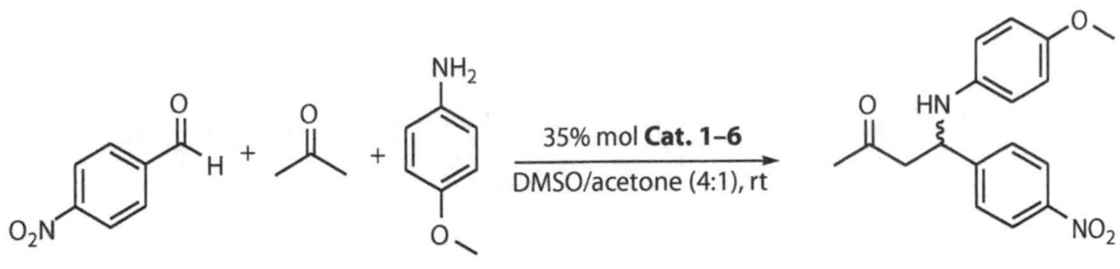

| 7 | 8 | 13 | | 14 |

Product (14). Compound **7** (50 mg, 0.33 mmol) was dissolved in 2 mL of DMSO. 1.1 equivalent of compound **13** (44 mg, 0.36 mmol) was added to the solution. 0.5 mL of acetone (**8**) was added to the mixture and stirred at room temperature for 10 min. 16.0 mg (0.12 mmol) of catalyst **1** was added. The reaction mixture was stirred to the completion and diluted with ethyl acetate (10 mL). The organic layer was washed twice with water (2 × 5 mL). The organic layer (product **14**) was dried over anhydrous Na_2SO_4 (2 g). According to the type of catalysts (**1–6**), 80–98 mg (77–95%) of product **14** was isolated (see Table 13.3). The reaction was repeated with catalysts **2–6** using the same mole ratio. ^1H NMR (400 MHz, 298 K, ppm, in $CDCl_3$): δ = 8.17 (d, J = 8.7 Hz, 2H, H-arom), 7.55 (d, J = 8.7 Hz, 2H, H-arom), 6.69 (d, J = 8.9 Hz, 2H, H-arom), 6.46 (d, J = 8.9 Hz, 2H, H-arom), 4.85 (t, J = 6.3 Hz, 1H, CH-NH), 4.23 (bs, 1H, NH), 3.69 (s, 3H, OCH_3), 2.95 (d, J = 6.3 Hz, 2H, $CO\text{-}CH_2$), 2.15 (s, 3H, $CO\text{-}CH_3$). ^{13}C NMR (101 MHz, 298 K, ppm, in $CDCl_3$): δ = 206.07 (C = O), 152.74 (C-arom), 150.63 (C-arom), 147.16 (C-arom), 140.10(C-arom), 127.38 (C-arom), 124.01 (2 × C-arom), 116.39 (C-arom), 115.36 (2 × C-arom), 114.77 (2 × C-arom), 55.60 (O-CH_3), 54.63 (N-CH), 50.65 (CO-CH_2), 30.66 (CO-CH_3).

TABLE 13.3
Summary of the Results of the Mannich Reaction

		Mannich Reaction		
Entry	Catalyst	Time of Reaction (day)	Conversion (%)	ee (%)
1	1	2	88	54
2	2	2	87	20
3	3	2	79	42
4	4	2	77	61
5	5	2	95	75
6	6	2	94	27

Knoevenagel Reaction

15   **16**   **17**

Dimethyl 2-(3-methyl-butylidene)malonate (17). 172 μl (1.55 mmol) of 3-methylbutyraldehyde (**16**) was dissolved in 10 mL DMSO. The addition of 23 mg (0.2 mmol) **1** was followed. After 5 min, 459 μl (4.00 mmol) dimethyl malonate (**15**) was added. The mixture was stirred overnight at room temperature. The reaction mixture was diluted with ethyl acetate (50 mL) and washed twice with water (2 × 20 mL). The organic layer was dried over anhydrous Na_2SO_4 (10 g). No further purification was needed. 270 mg (1.35 mmol, 87%) of product dimethyl 2-(3-methyl-butylidene) malonate (**17**) was isolated. The reaction was carried out with catalysts **2–6** in the same mole ratio. A yield of 85% to 90% was obtained. ^1H NMR (400 MHz, 298 K, ppm, in $CDCl_3$): δ = 7.07 (t, J = 7.9 Hz, 1H, = CH), 3.84 (s, 3H, OCH_3), 3.80 (s, 3H, OCH_3), 2.21 (dd, J = 6.9 Hz, J = 7.8 Hz, 2H, CH_2), 1.83 (m, 1H, $CH(CH_3)_2$), 0.96 (s, 3H, CH_3), 0.95 (s, 3H, CH_3).

GREEN METRICS ANALYSIS

In order to compare the material efficiency performances of the described methods with literature procedures to the same target molecules, a green metrics analysis was carried out.[34] Tables 13.4 to 13.6 summarize the essential material efficiency variables for the syntheses of the *R*-isomer, *S*-isomer, and combined stereoisomers of compound **9**, respectively. Similarly, Tables 13.7 to 13.9 summarize the essential material efficiency variables for the syntheses of the *R*-isomer, *S*-isomer, and combined stereoisomers of compound **12**, respectively. Data entries are ranked in ascending order of process mass intensity (PMI), where the most material-efficient plan appears at the top of the ranking. Schemes 13.1 and 13.2 summarize the organocatalysts used in the syntheses of stereoisomers of compounds **9** and **12**, respectively. Though no excess reagents were used in the aldol condensation to make compound **9** in quantitative yield, the reaction used a significant amount of auxiliary materials in the workup procedure, which ultimately reduced its overall PMI performance relative to other literature plans (see Table 13.6). Further improvements would clearly involve reducing the E-aux contribution to E-total while maintaining a high reaction yield using L-proline as catalyst. On the other hand, the Michael addition reaction to make compound **12** using catalyst **5** ranks second in performance behind Vaccaro's method using Rasta-TBD as catalyst (see Table 13.9).

TABLE 13.4
Green Metrics Summary for the Synthesis of the *R*-Isomer of Compound 9

Plan	Method[a]	Catalyst	% AE[b]	% Yield	E-Kernel	E-Excess	E-Aux	E-Total	PMI[c]
Fotaras (2012)[d]	A	C9; 4-nitro-benzoic acid	100	100	0.005	2.5	0.4	2.9	3.9
An (2010)[e]	A	C12	100	96	0.04	11.6	0.1	11.7	12.7
Li (2009)[f]—aqueous conditions	A	C3; $H_3PW_{12}O_{40}$	100	83	0.2	0.3	87.7	88.3	89.3
Liebscher (2012)[g]	A	Proline derivative bound to iron nanoparticle; benzoic acid	100	94	0.06	7.7	122.5	130.2	131.2
Li (2009)[f]—neat conditions	A	C3; $H_3PW_{12}O_{40}$	100	83	0.2	4.2	163.0	167.4	168.4
Whiting (2012)[h]	A	C4; (R,R)-hydrobenzoin; E_3N	100	80	0.2	3.1	283.8	287.2	288.2
Robak (2011)[i]	A	C11	100	81	0.2	9.9	396.4	406.5	407.5
Jia (2009)[j]	A	C13; 2,4-dinitrophenol	100	93	0.08	2.7	574.0	576.7	577.7
Sonoike (2012)[k]	A	C1-L	100	95	0.05	2.9	1784.0	1786.9	1787.9

[a] Method A = acetone + p-nitrobenzaldehyde aldol condensation.
[b] AE = atom economy.
[c] PMI = process mass intensity.
[d] Fotaras, S., Kokotos, C.G., Kokotos, G., *Org. Biomol. Chem.*, 2012, 10, 5613.
[e] An, Y.J., Zhang, Y.X., Wu, Y., Liu, Z.M., Pi, C., Tao, J.C., *Tetrahedron Asymm.*, 2010, 21, 688.
[f] Li, J., Hu, S., Luo, S., Cheng, J.P., *Eur. J. Org. Chem.*, 2009, 132.
[g] Yacob, Z., Nan, A., Liebscher, J., *Appl. Synth. Catal.*, 2012, 354, 3259.
[h] Georgiou, I., Whiting, A., *Eur. J. Org. Chem.*, 2012, 4110.
[i] Robak, M.T., Herbage, M.A., Ellman, J.A., *Tetrahedron*, 2011, 67, 4412.
[j] Jia, Y.N., Wu, F.C., Ma, X., Zhu, G.J., Da, C.S., *Tetrahedron Lett.*, 2009, 50, 3059.
[k] Sonoike, S., Itakura, T., Kitamura, M., Aoki, S., *Chem. Asian J.*, 2012, 7, 64.

TABLE 13.5

Green Metrics Summary for the Synthesis of the S-Isomer of Compound 9

Plan	Method[a]	Catalyst	% AE[b]	% Yield	E-Kernel	E-Excess	E-Aux	E-Total	PMI[c]
Arnold (2008)[d]	A	C2; Et$_3$N (R,R)-diisopropyltartrate	100	86	0.2	43.9	188.5	232.5	233.5
Sonoike (2012)[e]	A	C1-D	100	95	0.05	2.9	1784.0	1786.9	1787.9

[a] Method A = acetone + p-nitrobenzaldehyde aldol condensation.

[b] AE = atom economy.

[c] PMI = process mass intensity.

[d] Arnold, K., Batsanov, A.S., Davies, B., Grosjean, C., Schutz, T., Whiting, A., Zawatzky, K., *Chem. Commun.*, 2008, 3879.

[e] Sonoike, S., Itakura, T., Kitamura, M., Aoki, S., *Chem. Asian J.*, 2012, 7, 64.

TABLE 13.6

Green Metrics Summary for the Synthesis of Both Stereoisomers of Compound 9

Plan	Method[a]	Catalyst	% AE[b]	% Yield	E-Kernel	E-Excess	E-Aux	E-Total	PMI[c]
Kumar (2011)[d]	B	Novozym 435 Et$_3$N	67	89	0.7	0.003	0.01	0.7	1.7
Schaus (2004)[e]	E	C6; YbCl$_3$; Pd$_2$(dba)$_3$	67	76	1.0	13.9	17.5	32.3	33.3
Han (2007)[f]	C	Choline proline ionic liquid	100	90	0.1	2.8	35.2	38.1	39.1
Luo (2007)[g]	C	C10; acetic acid	100	88	0.1	2.8	77.4	80.4	81.4
Simpura (2000)[h]	A	Me$_3$Al Biphenyl-2,2′-diol	78	58	1.2	0	79.6	80.8	81.8
Yu (2011)[i]	C	SO3H-MCM-41-NH2	100	80	0.3	93.8	1	95.0	96.0
Banon-Caballero (2012)[j]	C	C5; benzoic acid	100	83	0.2	0.3	127.3	127.9	128.9
Reddy (2006)[k]	C	Chitosan hydrogel beads	100	87	0.1	4.0	129.2	133.4	134.4
Liu (2011)[l]	C	2-Naphthol; sodium 2-naphthoate	100	94	0.06	19.8	115.9	135.7	136.7
Schneider (2001)[m]	A	Zr(OBu)$_4$	78	85	0.5	0.7	152.2	153.3	154.3
Gryko (2007)[n]	A	C8; CF$_3$COOH	100	81	0.2	9.0	233.6	242.7	243.7
Whiting (2012)[o]	C	Pyrrolidine	100	99	0.01	37.9	206.4	244.3	245.3
Feng (2009)[p]	D	Candida antarctica lipase B; cyclen	70	96	0.5	250.0	20.1	270.6	271.6
Buonara (1995)[q]	C	Zn(NO$_3$)$_2$, 6 H$_2$O; N,N-dimethyl-ethanolamine	100	100	0	11.2	263.0	274.2	275.2
Chinni (2005)[r]	C	Pyrrolidine	100	91	0.1	5.8	473.3	479.2	480.2
Al-Momani (this work)	C	L-Proline	100	100	0	7.3	507.6	514.9	515.9
Tang (2005)[s]	C	C7	100	62	0.6	11.7	515.2	527.6	528.6

Continued

TABLE 13.6 (Continued)
Green Metrics Summary for the Synthesis of Both Stereoisomers of Compound 9

Plan	Method[a]	Catalyst	% AE[b]	% Yield	E-Kernel	E-Excess	E-Aux	E-Total	PMI[c]
Singh (2009)[t]	C	L-Proline	100	72	0.4	130.8	1729.2	1860.4	1861.4
Kubota (2003)[u]	C	Piperazine; FSM-16 (mesoporous silicate)	100	91	0.09	20.4	2228.1	2248.6	2249.6

[a] Method A = acetone + p-nitrobenzaldehyde aldol condensation, method B = isopropenyl acetate + p-nitrobenzaldehyde + isopropanol, method C = acetone + p-nitrobenzaldehyde aldol condensation, method D = ethyl acetoacetate + p-nitrobenzaldehyde decarboxylative aldol condensation, method E = allyl acetoacetate + p-nitrobenzaldehyde decarboxylative aldol condensation.

[b] AE = atom economy.

[c] PMI = process mass intensity.

[d] Kumar, M., Shah, B.A., Taneja, S.C., *Appl. Synth. Catal.*, 2011, 353, 1207.

[e] Lou, S., Westbrook, J., Schaus, S.E., *J. Am. Chem. Soc.*, 2004, 126, 11440.

[f] Hu, S., Jiang, T., Zhang, Z., Zhu, A., Han, B., Song, J., Xie, Li, W., *Tetrahedron Lett.*, 2007, 48, 5613.

[g] Luo, S., Mi, X., Zhang, L., Liu, S., Xu, H., Cheng, J.P., *Tetrahedron*, 2007, 63, 1923.

[h] Simpura, I., Nevalainen, V., *Angew. Chem. Int. Ed.*, 2000, 39, 3422.

[i] Yu, X., Zou, Y., Wu, S., Liu, H., Guan, J., Kan, Q., *Mater. Res. Bull.*, 2011, 46, 951.

[j] Banon-Caballero, A., Guillena, G., Najera, C., *Helv. Chim. Acta*, 2012, 95, 1831.

[k] Reddy, K.R., Rajgopal, K., Maheswari, C.U., Kantam, M.L., *New J. Chem.*, 2006, 30, 1549.

[l] Liu, C.L., Zhang, H., Wang, N., *Chin. Chem. Lett.*, 2011, 22, 651.

[m] Schneider, C., Hansch, M., *Chem. Commun.*, 2001, 1218.

[n] Gryko, D., Zimnicka, M., Lipinski, R., *J. Org. Chem.*, 2007, 72, 964.

[o] Georgiou, I., Whiting, A., *Org. Biomol. Chem.*, 2012, 10, 2422.

[p] Feng, X.W., Li, C., Wang, N., Li, K., Zhang, W.W., Wang, Z., Yu, X.Q., *Green Chem.*, 2009, 11, 1933.

[q] Buonara, P.T., Rosauer, K.G., Dai, L., *Tetrahedron Lett.*, 1995, 36, 4009.

[r] Chimni, S.S., Mahajan, D., *Tetrahedron*, 2005, 61, 5019.

[s] Tang, Z., Yang, Z.H., Chen, X.H., Cun, L.F., Mi, A.Q., Jiang, Y.Z., Gong, L.Z., *J. Am. Chem. Soc.*, 2005, 127, 9285.

[t] Singh, P., Bhardwaj, A., Kaur, S., Kumar, S., *Eur. J. Med. Chem.*, 2009, 44, 1278.

[u] Kubota, Y., Goto, K., Miyata, S., Goto, Y., Fukushima, Y., Sugi, Y., *Chem. Lett.*, 2003, 32, 234.

TABLE 13.7

Green Metrics Summary for the Synthesis of the *R*-Isomer of Compound 12

Plan	Method[a]	Catalyst	% AE[b]	% Yield	E-Kernel	E-Excess	E-Aux	E-Total	PMI[c]
Vaccaro (2012)[d]	A	C16	100	97	0.03	0.001	7.4	7.4	8.4
Kawara (1994)[e]	A	C20	100	53	0.9	0	11.5	>12.3	>13.3
Jorgensen (2003)[f]	A	C14	100	66	0.5	11.8	0.1	>12.5	>13.5
Yang (2008)[g]	A	C15	100	99	0.007	0.5	95.1	95.6	96.6
Luo (2012)[h]	A	C17	100	98	0.02	0.5	>160.0	>160.5	>160.5
Riguet (2009)[i]	A	C21	100	94	0.07	75.6	>1278.6	>1354.3	>1355.3

[a] Method A = chalcone + dimethyl malonate Michael addition.

[b] AE = atom economy.

[c] PMI = process mass intensity.

[d] Bonollo, S., Lanari, D., Longo, J.M., Vaccaro, L., *Green Chem.*, 2012, 14, 164.

[e] Kawara, A., Taguchi, T., *Tetrahedron Lett.*, 1994, 35, 8805.

[f] Halland, N., Aburel, P.S., Jorgensen, K.A., *Angew. Chem. Int. Ed.*, 2003, 42, 661.

[g] Yang, Y.Q., Zhao, G., *Chem. Eur. J.*, 2008, 14, 10888.

[h] Luo, C., Jin, Y., Du, D.M., *Org. Biomol. Chem.*, 2012, 10, 4116.

[i] Riguet, E., *Tetrahedron Lett.*, 2009, 50, 4283.

TABLE 13.8
Green Metrics Summary for the Synthesis of the S-Isomer of Compound 12

Plan	Method[a]	Catalyst	% AE[b]	% Yield	E-Kernel	E-Excess	E-Aux	E-Total	PMI[c]
Li (2009)[d]	B	C18	100	92	0.09	1.0	46.2	47.3	48.3
Dudzinski (2012)[e]	A	C19	100	85	0.2	0.1	54.6	54.9	55.9
Jacobsen (2005)[f]	A, step 1	(R,R)-[(salen)Al]2O	100	91	0.1	0.09	121.1	121.3	122.3
	A, step 2	None	73	76	0.8	41.9	1620.8	1663.5	1664.5
Jacobsen (2005)[f]	A, combined		73	69	0.9	41.9	1767.3	1810.2	1811.2

[a] Method A = step 1: chalcone + methyl cyanoacetate Michael addition, step 2: nitrile hydrolysis, method B = chalcone + dimethyl malonate Michael addition.

[b] AE = atom economy.

[c] PMI = process mass intensity.

[d] Li, P., Wen, S., Yu, F., Liu, Q., Li, W., Wang, Y., Liang, X., Ye, J., *Org. Lett.*, 2009, 11, 753.

[e] Dudzinski, K., Pakulska, A.M., Kwiatkowski, P., *Org. Lett.*, 2012, 14, 4222.

[f] Taylor, M.S., Zalatan, D.N., Lerchner, A.M., Jacobsen, E.N., *J. Am. Chem. Soc.*, 2005, 127, 1313.

TABLE 13.9

Green Metrics Summary for the Synthesis of Both Stereoisomers of Compound 12

Plan	Method[a]	Catalyst	% AE[b]	% Yield	E-Kernel	E-Excess	E-Aux	E-Total	PMI[c]
Vaccaro (2012)[e]	A	Rasta-TBD[d]	100	92	0.09	0.001	17.9	18.0	19.0
Al-Momani (2012,[f] this work)	A	Catalyst 5	100	76	0.3	0.4	113.8	114.5	115.5
Sukanya (2011)[g]	A	LiOH	100	88	0.1	0	981.9	982.0	983.0

[a] Method A = chalcone + dimethyl malonate.

[b] AE = atom economy.

[c] PMI = process mass intensity.

[d] TBD = 1,5,7-Triazabicyclo[4.4.0]dec-5-ene.

[e] Bonollo, S., Lanari, D., Angelini, T., Pizzo, F., Marrocchi, A., Vaccaro, L., *J. Catal.*, 2012, 285, 216.

[f] Al-Momani, L.A., *ARKIVOC* 2012, vi, 101.

[g] Sukanya, K., Deka, D.C., *Ind. J. Chem.*, 2011, 50B, 872.

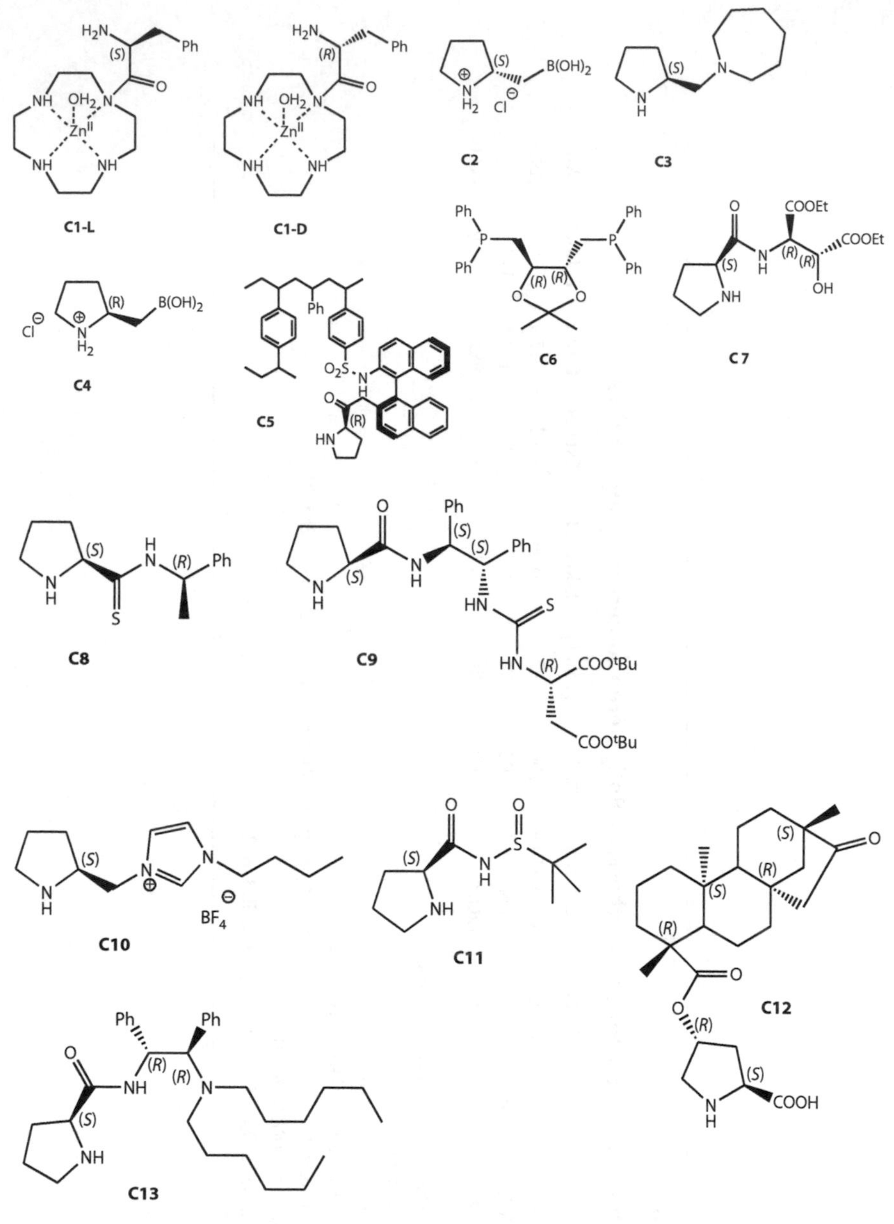

SCHEME 13.1 Organocatalysts used to synthesize compound **9** (see Tables 13.4 to 13.6).

For compound **14** Tables 13.10 and 13.11 summarize the respective green metrics performances for the syntheses of the *S*-isomer and the combined isomers. Scheme 13.3 shows the structures of the associated organocatalysts used. The Mannich reaction using catalyst **4** ranks second behind Sueki's method using an iridium catalyst (see Table 13.10).

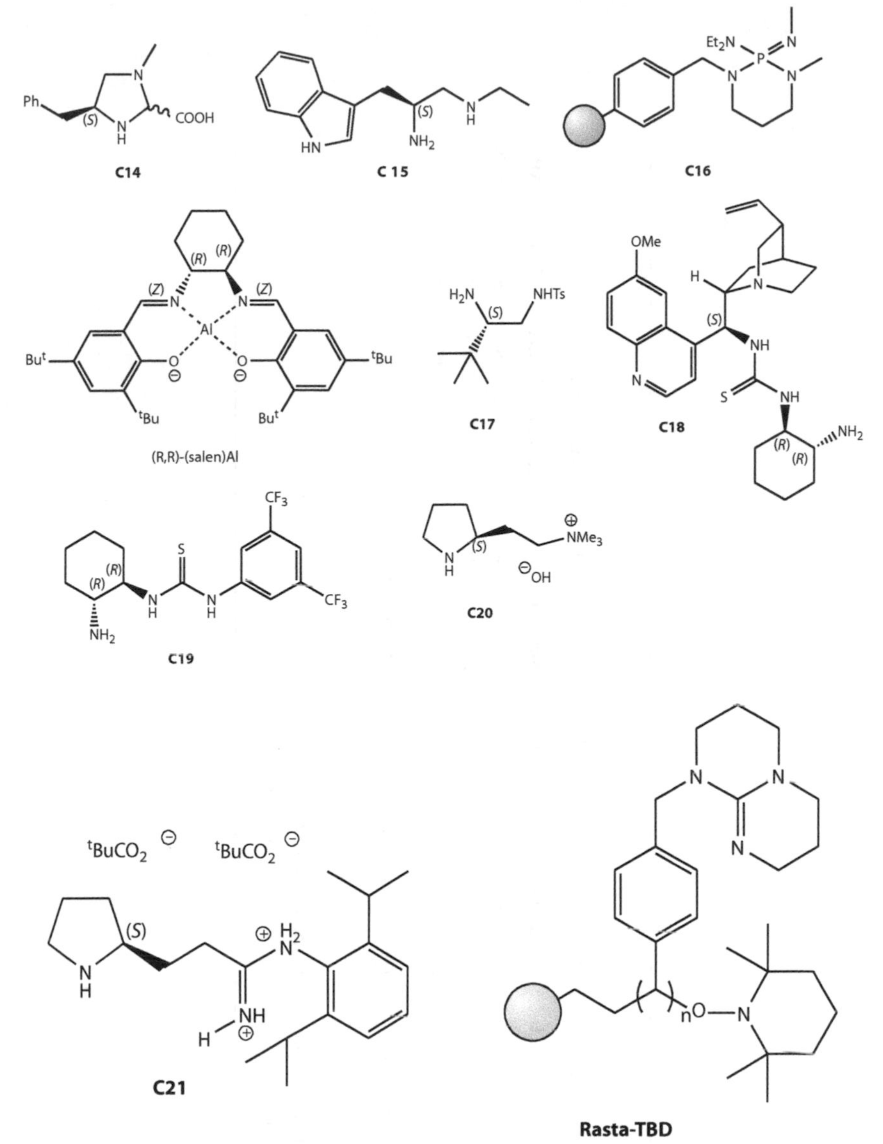

SCHEME 13.2 Organocatalysts used to synthesize compound **12** (see Tables 13.7 to 13.9).

Table 13.12 summarizes the green metrics performances of various methods to synthesize compound **17**. The Knoevenagel reaction in this work using ʟ-proline as catalyst ranks last in material efficiency due to the high consumption of auxiliary materials in the workup and purification procedures. Further improvements to reduce solvent demand would make the present method competitive with previously described methods. Interestingly, the oldest method due to Stork using piperidine as catalyst still ranks as the most material-efficient method to make this compound (see Table 13.12).

TABLE 13.10
Green Metrics Summary for the Synthesis of Both Stereoisomers of Compound 14

Plan	Method[a]	Catalyst	% AE[b]	% Yield	E-Kernel	E-Excess	E-Aux	E-Total	PMI[c]
Sueki (2006)[d]	A	$[IrCl_2(H)(cod)]_2$	95	57	0.9	26.4	>25.0	>52.3	>53.3
Al-Momani (this work)	A	Catalyst 4	95	76	0.4	4.7	290.4	295.5	296.5
Chai (2010)[e]	A	Trypsin	95	93	0.1	13.4	390.0	403.5	404.5

[a] Method A = acetone + p-methoxyaniline + p-nitrobenzaldehyde Mannich condensation.

[b] AE = atom economy.

[c] PMI = process mass intensity.

[d] Sueki, S., Igarashi, T., Nakajima, T., Shimizu, I., *Chem. Lett.*, 2006, 35, 682.

[e] Chai, S.J., Lai, Y.F., Zheng, H., Zhang, P.F., *Helv. Chim. Acta*, 2010, 93, 2231.

TABLE 13.11
Green Metrics Summary for the Synthesis of the S-Isomer of Compound 14

Plan	Method[a]	Catalyst	% AE[b]	% Yield	E-Kernel	E-Excess	E-Aux	E-Total	PMI[c]
Notz (2001)[d]	B	C23	95	52	1.0	4.5	27.1	>32.6	>33.6
List (2002)[e]	B	S-Proline	95	50	1.1	9.8	1025.1	1036.0	1037.0
Barbas (2003)[f]	B	S-Proline	95	54	1.0	8.2	1533.0	1542.2	1543.2
Hayashi (2003)[g]	B	S-Proline	95	58	0.8	8.4	4992.1	5001.3	5002.3
Benaglia (2002)[h]	A	C22	100	80	0.3	15.6	8843.5	8859.3	8860.3

[a] Method A = acetone + N-4-methoxyphenyl-4-nitrobenzaldimine, method B = acetone + p-methoxyaniline + p-nitrobenzaldehyde Mannich condensation.

[b] AE = atom economy.

[c] PMI = process mass intensity.

[d] Notz, W., Sakthivel, K., Bui, T., Zhong, G., Barbas, C.F., III, *Tetrahedron Lett.*, 2001, 42, 199.

[e] List, B., Pojarliev, P., Biller, W.T., Martin, H.J., *J. Am. Chem. Soc.*, 2002, 124, 827.

[f] Chowdari, N.S., Ramachary, D.B., Barbas, C.F., III, *Synlett*, 2003, 1906.

[g] Hayashi, Y., Tsuboi, W., Shoji, M., Suzuki, N., *J. Am. Chem. Soc.*, 2003, 125, 11208.

[h] Benaglia, M., Cinquini, M., Cozzi, F., Puglisi, A., Celentano, G., *Adv. Synth. Catal.*, 2002, 344, 533.

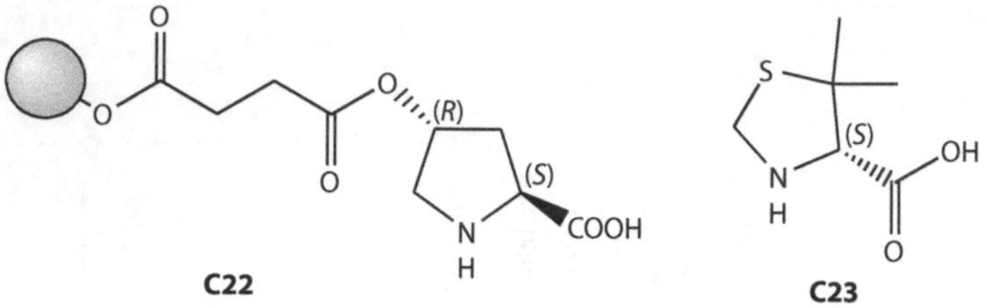

C22   **C23**

SCHEME 13.3   Organocatalysts used to synthesize compound **14** (see Tables 13.10 and 13.11).

TABLE 13.12
Green Metrics Summary for the Synthesis of Compound 17

Plan	Method[a]	Catalyst	% AE[b]	% Yield	E-Kernel	E-Excess	E-Aux	E-Total	PMI[c]
Stork (1974)[d]	A	Piperidine	92	83	0.3	0.004	8.7	9.0	10.0
Synthon (2009)[e]	A	L-Proline	92	88	0.2	0.7	11.4	12.4	13.4
Cardillo (2003)[f]	A	L-Proline	92	92	0.2	0.7	13.3	14.2	15.2
Biediger (2004)[g]	A	Piperidine	92	68	0.6	0.01	14.8	15.5	16.5
Zhou (1993)[h]	C	CuI	41	88	1.8	0.5	57.1	59.4	60.4
Zhou (1991)[i]	B	None	23	91	3.7	0	98.4	102.1	103.1
DSM (2005)[j]	A	L-Proline	92	96	0.1	1.1	141.6	142.8	143.8
Al-Momani (this work)	A	L-Proline	92	87	0.3	1.2	393.1	394.5	395.5

[a] Method A = isovaleraldehyde + dimethyl malonate condensation, method B = isovaleraldehyde + dimethyl 2,2-dibromomalonate + dibutyl telluride, method C = isovaleraldehyde + dimethyl 2-diazomalonate + dibutyl telluride.

[b] AE = atom economy.

[c] PMI = process mass intensity.

[d] Stork, G., Szajewski, R.P., *J. Am. Chem. Soc.,* 1974, 96, 5787.

[e] Thijs, L., U.S. 2009312560 (Synthon IP, Inc., 2009).

[f] Cardillo, G., Fabbroni, S., Gentilucci, L., Gianotti, M., Tolomelli, A., *Synth. Commun.,* 2003, 33, 1587.

[g] Biediger, R.J., Chen, Q., Decker, E.R., Holland, G.W., Kassir, J.M., Li, W., Market, R.V., Scott, I.L., Li, J., U.S. 2004063955 (2004).

[h] Zhou, Z.L., Huang, Y.Z., Shi, L.L., *Tetrahedron,* 1993, 49, 6821.

[i] Zhou, Z.L., Shi, L.L., Huang, Y.Z., *Synth. Commun.,* 1991, 21, 1027.

[j] Wubbolts, M.G., Bovenberg, R.A., Sprenger, G., Bongaerts, J.J., Kozak, S., Muller, M., EP 1728872 (DSM, 2005).

REFERENCES

1. A. Berkessel and H. Gröger, *Asymmetric Organocatalysis: From Biomimetic Concepts to Applications in Asymmetric Synthesis*, Wiley-VCH, Weinheim, 2005, pp. 45–165.
2. U. Eder, G. Sauer, and R. Wiechert, *Angew. Chem. Int. Ed.* 1971, 10, 496–497.
3. Z.G. Hajos and D.R. Parrish, *J. Org. Chem.* 1974, 39, 1615–1621.
4. E.R. Jarvo and S.T. Miller, *Tetrahedron* 2002, 58, 2481–2495.
5. L. Al-Momani, *ARKIVOC* 2012, vi, 101–111.
6. L. Al-Momani and A. Lataifeh, *Inorg. Chim. Acta* 2013, 394, 176–183.
7. A. Cordova, W. Notz, G. Zhong, J.M. Betancort, and C.F. Barbas III, *J. Am. Chem. Soc.* 2002, 124, 1842–1843.
8. L. Hoang, S. Bahmanyar, K.N. Houk, and B. List, *J. Am. Chem. Soc.* 2003, 125, 16–17.
9. B. List, R.A. Lerner, and C.F. Barbas III, *J. Am. Chem. Soc.* 2000, 122, 2395–2396.
10. W. Notz, F. Tanaka, and C.F. Barbas III, *Acc. Chem. Res.* 2004, 37, 580–591.
11. M.B. Schmid, K. Zeitler, and R.M. Gschwind, *Angew. Chem. Int. Ed.* 2010, 49, 4997–5003.
12. C. Klein and W. Hüttel, *Adv. Synth. Catal.* 2011, 353, 1375–1383.
13. P.-Q. Huang and H.-Y. Huang, *Synth. Commun.* 2004, 34, 1377–1382.
14. A. Hartikka and P.I. Arvidsson, *Tetrahedron Asymm.* 2004, 15, 1831.
15. A. Hartikka and P.I. Arvidsson, *Eur. J. Org. Chem.* 2005, 4287–4295.
16. M. Raj, S.K.G. Vishnumaya, and V.K. Singh, *Org. Lett.* 2006, 8, 4097–4099.
17. Y. Hayashi, T. Sumiya, J. Takahashi, H. Gotoh, T. Urushima, and M. Shoji, *Angew. Chem. Int. Ed.* 2006, 45, 958–961.
18. S.-P. Zhang, X.-K. Fu, and S.-D. Fu, *Tetrahedron Lett.* 2009, 50, 1173–1176.
19. Z. Tang, F. Jiang, L.T. Yu, X. Cui, L.Z. Gong, A.Q. Mi, Y.Z. Jiang, and Y.D. Wu, *J. Am. Chem. Soc.* 2003, 125, 5262–5263.
20. A. Berkessel, A. Koch, and J. Lex, *Adv. Synth. Catal.* 2004, 346, 1141–1146.
21. S.S. Chimni and D. Mahajan, *Tetrahedron Lett.* 2005, 46, 5617–5619.
22. S.S. Chimni and D. Mahajan, *Tetrahedron Asymm.* 2006, 17, 2108–2119.
23. Z. Tang, Z.H. Yang, X.H. Chen, L.F. Cun, A.Q. Mi, Y.Z. Jiang, and L.Z. Gong, *J. Am. Chem. Soc.* 2005, 127, 9285–9289.
24. D. Gryko and R. Lipinski, *Adv. Synth. Catal.* 2005, 347, 1948–1952.
25. D. Gryko and R. Lipinski, *Eur. J. Org. Chem.* 2006, 3864–3876.
26. K. Sakthivel, W. Notz, T. Bui, and C.F. Barbas III, *J. Am. Chem. Soc.* 2001, 123, 5260–5267.
27. R. Ait-Youcef, D. Kalch, X. Moreau, C. Thomassigny, and C. Greck, *Lett. Org. Chem.* 2009, 6, 377–380.
28. J. Kofoed, T. Darbre, and J.-L. Reymond, *Chem. Commun.* 2006, 1482–1484.
29. J. Paradowska, M. Stodulski, and J. Mlynarski, *Angew. Chem. Int. Ed.* 2009, 48, 4288–4297.
30. S. Itoh, M. Kitamura, Y. Yasuyuki, and S. Aoki, *Chem. Eur. J.* 2009, 15, 10570–10584.
31. R. Hara and K. Kino, *Biochem. Biophys. Res. Commun.* 2009, 379, 882–886.
32. H. Mori, T. Shibasaki, K. Yano, and A. Ozaki, *J. Bacteriol.* 1997, 179, 5677–5683.
33. L. Petersen, R. Olewinski, P. Salmon, and N. Connors, *Appl. Microbiol. Biotechnol.* 2003, 62, 263–267.
34. (a) C. Jiménez-González, C.S. Ponder, Q.B. Broxterman, and J.B. Manley, *Org. Proc. Res. Dev.* 2011, 15, 912–917; (b) J. Andraos, *Org. Process Res. Dev.* 2009, 13, 161–185.

14 Green Synthesis of Homoallylic Silyl Ethers

Matthew R. Dintzner

CONTENTS

INTRODUCTION

Allylation of carbonyl compounds (**1**) is arguably one of the most important and widely used carbon-carbon bond-forming reactions in organic synthesis (Scheme 14.1).[1] The resulting homoallylic alcohols (**2**) are useful platforms for further synthetic elaboration, especially in the event that the allylation reaction proceeds stereoselectively.[2] Accordingly, significant effort over the past several decades has been aimed at the development of synthetic methodologies for effecting this carbon-carbon bond-forming reaction, including those that employ organometallic reagents like Grignard reagents, allylboranes, and allylsilanes.[1] The latter, a Lewis acid-catalyzed reaction of a carbonyl with an allylsilane reagent, is known as the Hosomi–Sakurai or Sakurai reaction and has been extensively studied and applied in natural products synthesis.[3]

Although conventional Lewis acids, like $SnCl_4$, $TiCl_4$, $BF_3 \cdot OEt_2$, $AlCl_3$, and $Sc(OTf)_3$, have been shown to successfully catalyze the Hosomi–Sakurai reaction, the negative impact of these reagents on the environment has encouraged the development of alternative, greener methodologies for effecting this and related synthetic transformations.[4] Toward that end, a more environmentally benign procedure for the Hosomi–Sakurai reaction was investigated.[5] It was observed that activated montmorillonite K10 clay is an extremely efficient catalyst for the addition of allyltrimethylsilane to electron-deficient benzaldehydes (**3**) to give homoallylic silyl ethers (**4**) in excellent yields (Scheme 14.2).[5] The reaction proceeds rapidly at room temperature or below in the presence of minimal organic solvent with perfect atom economy.

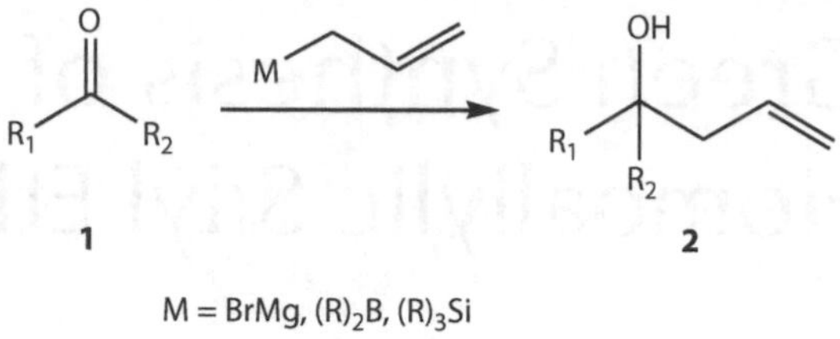

$$M = BrMg, (R)_2B, (R)_3Si$$

SCHEME 14.1   Allylation of carbonyls.

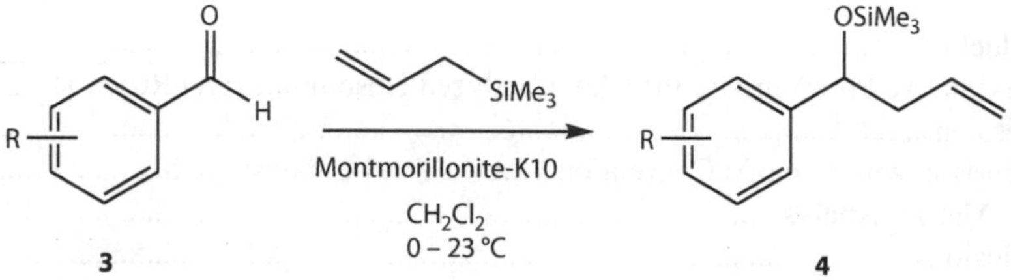

SCHEME 14.2   Clay-catalyzed Hosomi–Sakurai reaction.

In the most general sense, clays are a type of fine-grained earth, primarily composed of extensively layered aluminum and silicate minerals.[6] Montmorillonite clays are thought to have formed from volcanic ash during the Jurassic and later periods and were named for the location of their discovery, Montmorillon, France, in the 1800s. Montmorillonite clays are now mined from regions all over the world, including Europe, Africa, Asia, and South and North America. Montmorillonite K10, which is a member of the smectite family of minerals, has a layered structure that consists of sheets of tetrahedral silicate (SiO_4^{4-}) ions that sandwich sheets of octahedral aluminum (Al^{3+}) oxide (T-O-T scaffold, Figure 14.1), forming an interconnected network via weak bonds with oxygen. Intermittent substitutions of Al^{3+} or Fe^{3+} for Si^{4+}, or Fe^{2+} or Mg^{2+} for Al^{3+} (M, Figure 14.1), result in a charge imbalance that is compensated for by intercalation of alkaline earth metal cations like Na^+ and Mg^{2+}, and water, in the spaces between the layers (interlayer region, Figure 14.1). These structural features impart some Lewis and Bronsted acidic properties to the clay that may be exploited in catalysis. In addition, acid-treated clays, such as montmorillonite K10, are available through manufacturers like Sigma-Aldrich in the United States and have been used extensively in organic synthesis.[6,7]

Previous work[8] in our laboratories has shown that montmorillonite K10 clay is effective at activating aromatic aldehydes toward nucleophilic attack, particularly when the clay is first activated by heating (200°C, 1 h).[9] Heating results in collapse of the clay's interlayer structure and marked increase in its Lewis acidity, as water is extruded.[9] In the present work, we report the clay-catalyzed reaction of allyltrimethylsilane with aromatic aldehydes to give homoallylic silyl ethers.[5]

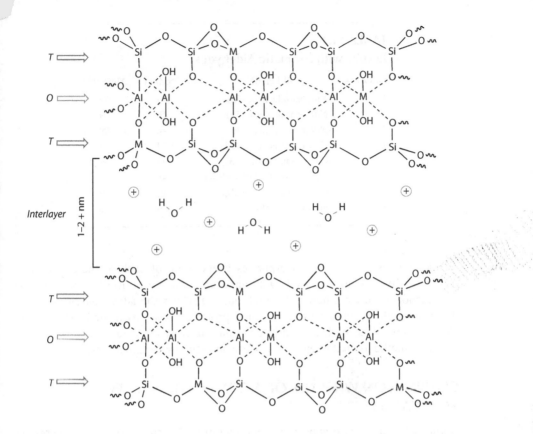

FIGURE 14.1 Layered structure of montmorillonite clays (T = tetrahedral, O = octahedral).

REPRESENTATIVE PROCEDURE FOR THE CLAY-CATALYZED HOSOMI–SAKURAI REACTION

Montmorillonite K10 clay (200 mg) is added to a glass scintillation vial and heated in an oven at 200°C for 1 h. The clay is cooled to ambient temperature in a desiccator, and then treated with 4-nitrobenzaldehyde (200 mg, 1.32 mmol) and 1 mL dichloromethane. The mixture is cooled to 0°C in an ice water bath and stirred gently with a magnetic stir bar. To the cooled mixture is added allyltrimethylsilane (1.98 mmol), and the resulting mixture is stirred at 0°C for 5 min. The ice water bath is removed and the reaction mixture is allowed to warm to ambient temperature over the course of 1 h. The reaction mixture is vacuum filtered, washing with an additional 1 mL of dichloromethane, and concentrated under vacuum to give the desired product (**4c**) as a pale yellow oil (345 mg, 99%) requiring no further purification.[5]

DISCUSSION

The scope of the reaction was extended to a variety of benzaldehyde derivatives and the results presented in Table 14.1. Because both the catalyst and the solvent can be recovered and reused, this procedure generates minimal waste (E-factor = 0.23).

TABLE 14.1
Results with Aromatic Aldehydes

Entry	Aldehyde	Product	Yield (%)
1	2-Nitrobenzaldehyde (**3a**)	**4a**	84
2	3-Nitrobenzaldehyde (**3b**)	**4b**	99
3	4-Nitrobenzaldehyde (**3c**)	**4c**	99
4	3-Fluorobenzaldehyde (**3d**)	**4d**	97
5	4-Chlorobenzaldehyde (**3e**)	**4e**	99
6	2,6-Dichlorobenzaldehyde (**3f**)	**4f**	99
7	3-Cyanobenzaldehyde (**3g**)	**4g**	99
8	Benzaldehyde (**3h**)	**4h**	75

In addition, as the product incorporates every atom of each of the two organic reactants, this reaction proceeds with an atom economy of 100%. Not only does the reaction generate a new C-C bond, but it also has the advantage of providing the homoallylic alcohol products in their protected form, as trimethylsilyl ethers. Deprotection, if desired, is readily and quantitatively effected by adding a slight excess of methanol to the reaction mixture and stirring for 20 min prior to filtering.

COMPARISON WITH USE OF A CONVENTIONAL LEWIS ACID CATALYST

Numerous reports have appeared in the literature describing Hosomi–Sakurai and related reactions catalyzed by more traditional Lewis acids.[10] For example, Warner and coworkers reported the ytterbium trichloride-catalyzed allylation of aldehydes with allyltrimethylsilane.[11] While YbCl$_3$ was observed to be an effective catalyst for this reaction, it is nonetheless considerably toxic and expensive (more than $17 per gram from Sigma-Aldrich) relative to montmorillonite K10 clay (less than $0.10 per gram). In addition, the reaction was carried out in excess nitromethane, required considerably longer reaction times (approximately 16 h), and much more extensive workup and purification than our procedure. A green metrics analysis for the clay-catalyzed allylsilation of 4-nitrobenzaldehyde (**3c**) to give **4c** is presented in Table 14.2. While the YbCl$_3$-catalyzed reaction of **3c** also proceeded with perfect atom economy, the reported yield of **4c** was lower (90%), and the corresponding E-factor was considerably higher given the lower yield and the need for aqueous workup and drying steps (exact amounts of reagents used were not reported in the literature).[11]

GREEN METRICS ANALYSIS

In this section we compare the material efficiencies for our syntheses of compounds **4c** and **4h** with known literature procedures to the same targets.[12,13] Compound **4c** has been made via the Hosomi–Sakurai method but with Al-MCM-41 (mesoporous aluminosilicate) as catalyst. Compound **4h** has also been made with the same

TABLE 14.2

Green Metrics Analysis for the Clay-Catalyzed Reaction of 4-Nitrobenzaldehyde with Allyltrimethylsilane (99% Yield)

Input		Output	
4-Nitrobenzaldehyde (3c)	0.200 g	1-Nitro-4-[1-[(trimethylsilyl)oxy]-3-buten-1-yl]-	0.345 g
Allyltrimethylsilane	0.226 g	benzene (4c)	
Dichloromethane	2.660 g	Recovered solvent	2.660 g
Montmorillonite K10 clay	0.200 g	Recovered clay	0.200 g
Total (used)	3.286 g	Total waste (mass input – mass output)	0.081 g

E-factor = 0.23

Atom economy = 100%

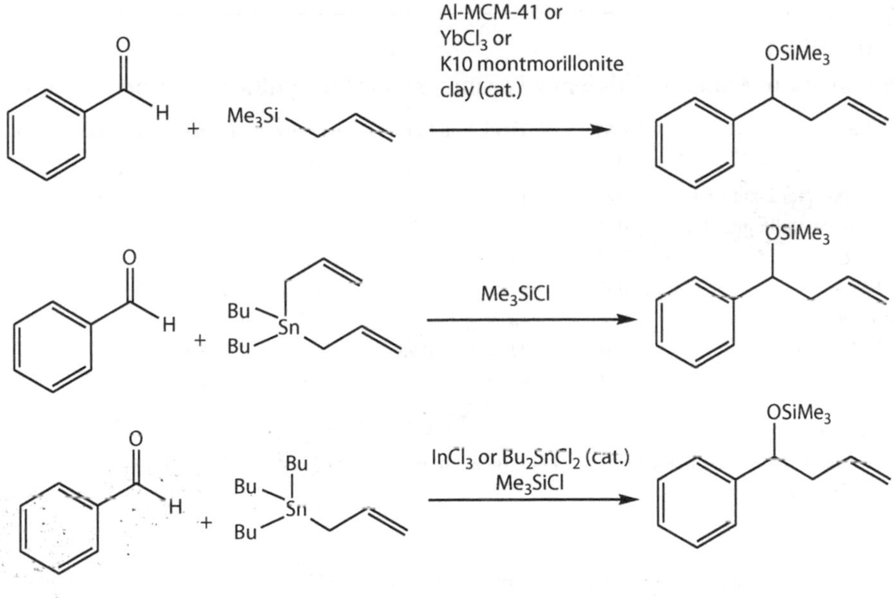

SCHEME 14.3   Various synthesis routes to compound **4h**.

methodology, and in addition by reactions shown in Scheme 14.3. Tables 14.3 and 14.4 summarize the relevant green metrics determined using an algorithm that partitions the waste contributions to the overall E-factor under the assumption that no materials except for the desired product are recovered.[14] Since literature plans did not disclose quantities of materials used in purification procedures, the auxiliary materials contribution to E-total, E-aux, takes into account reaction solvents, catalysts, and workup materials only. Plans are ranked according to ascending values of the process mass intensity (PMI).[15] The results of Table 14.3 show that the protocol described in this work ranks first for preparing compound **4c**. For compound

TABLE 14.3
Summary of Material Efficiency Parameters for the Synthesis of Compound 4c

Catalyst	% AE[a]	% Yield	E-Kernel	E-Excess	E-Aux[b]	E-Total	PMI
K10 montmorillonite clay[f]	100	98	0.02	0.2	8.3	8.6	9.6
K10 montmorillonite clay[e]	100	99	0.01	0.2	12.2	12.4	13.4
Al-MCM-41[d]	100	92	0.09	0.2	227.4	227.7	228.7
Al-MCM-41[c]	100	92	0.08	0.5	322.5	323.1	324.1

[a] AE = atom economy.
[b] Excludes chromatographic materials in purification procedures.
[c] See reference 12a.
[d] See reference 12b.
[e] See reference 5.
[f] This work.

TABLE 14.4
Summary of Material Efficiency Parameters for the Synthesis of Compound 4h

Catalyst	% AE[a]	% Yield	E-Kernel	E-Excess	E-Aux[b]	E-Total	PMI
Al-MCM-41[i]	100	92	0.08	0.3	13.3	13.7	14.7
K10 montmorillonite clay[j]	100	75	0.3	0.3	19.3	20.0	21.0
K10 montmorillonite clay[k]	100	75	0.3	0.3	19.3	20.0	21.0
None[d]	59.2	99	0.7	1.7	41.7	44.1	45.1
YbCl$_3$[h]	100	62	0.6	0.3	117.4	118.4	119.4
Bu$_2$SnCl$_2$[g]	40.4	76	2.3	0.2	126.9	129.4	130.4
Al-MCM-4[c]	100	92	0.08	0.3	273.3	273.6	274.6
InCl$_3$[e]	40.4	98	ND[f]	ND	ND	ND	ND

[a] AE = atom economy.
[b] Excludes chromatographic materials in purification procedures.
[c] See reference 12a.
[d] See reference 13b.
[e] See reference 13c.
[f] ND = not determinable since no quantities of materials were specified in the experimental procedure.
[g] See reference 13d.
[h] See reference 11.
[i] See reference 12b.
[j] See reference 5.
[k] This work.

4h the best performance using the Hosami-Sakurai methodology is found with Al-MCM-41 over the montmorillonite clay as catalyst since the reported yields are 92% and 75%, respectively. All other green metrics parameters between the two protocols are essentially the same. This difference in reaction yield is responsible for the second-place ranking. Other methodologies using allyltin reagents and

trimethylsilyl chloride rank lower in PMI because the respective atom economies are considerably lower due the production of by-products, and the auxiliary material consumptions were also higher.

CONCLUSIONS

In summary, we have found that activated montmorillonite K10 clay efficiently catalyzes the addition of allyltrimethylsilane to electron-deficient aromatic aldehydes at room temperature or below in dichloromethane to give homoallylic silyl ethers in good to excellent yields. This methodology constitutes a milder, more environmentally friendly, and more cost-effective alternative to the use of transition metal halides or oxides for the catalysis of the Hosomi–Sakurai reaction.

NOTES

Proton nuclear magnetic resonance (^{1}H) spectra and carbon-13 (^{13}C) spectra were recorded at 400 MHz and 100 MHz, respectively. The proton signal of residual, nondeuterated solvent ($\delta 7.26$ ppm for $CHCl_3$) was used as an internal reference for ^{1}H spectra. For ^{13}C spectra, chemical shifts are reported relative to the $\delta 77.23$ ppm resonance of $CDCl_3$. Coupling constants are reported in Hz. Infrared spectra were recorded as thin films on a Nicolet Avatar 360. Gas chromatographic analysis was performed on a Hewlett Packard 5890 Series II gas chromatograph with a 5971 Series mass selective detector.

1-Nitro-2-[1-[(trimethylsilyl)oxy]-3-buten-1-yl]benzene (4a): Infrared (IR) ($CDCl_3$) 3077, 2958, 1701, 1641, 1610, 1577, 1527, 1445, 1346, 1298, 1252 cm^{-1}; ^{1}H nuclear magnetic resonance (NMR) (300 MHz, $CDCl_3$) δ 7.89 (m, 2H), 7.63 (m, 1H), 7.40 (m, 1H), 5.86 (m, 1H), 5.38 (dd, J = 4.05, 7.63 Hz), 5.07 (m, 2H), 2.52, (m, 1H), 2.41 (m, 1H), 0.05 (s, 9H); ^{13}C NMR (75 MHz, $CDCl_3$) δ 147.17, 140.64, 139.31, 134.46, 133.12, 128.63, 127.72, 123.98, 117.59, 69.63, 44.24, 0.27; gas chromatography–mass spectrometry (GC-MS) (70 eV), t_R = 11.843 min, m/z 265 M$^+$ (<1%); 224 [M-41]$^+$ (60%); 73 [M-192]$^+$ (100%).

1-Nitro-3-[1-[(trimethylsilyl)oxy]-3-buten-1-yl]benzene (4b): IR ($CDCl_3$) 3077, 2958, 2902, 1641, 1432, 1479, 1437, 1351, 1310, 1252 cm^{-1}; ^{1}H NMR (300 MHz, $CDCl_3$) δ 8.19 (s, 1H), 8.10 (ddd, J = 1.13, 2.26, 8.1 Hz, 1H), 7.66 (d, J = 7.72 Hz, 1H), 7.49 (t, J = 7.91 Hz, 1H), 5.75 (m, 1H), 5.02 (m, 2H), 4.81 (dd, J = 5.56, 6.88 Hz, 1H), 2.45 (m, 2H), 0.07 (s, 9H); ^{13}C NMR (75 MHz, $CDCl_3$) δ 148.21, 147.12, 133.96, 131.99, 129.06, 122.12, 120.89, 117.88, 73.75, 44.88, 0.05; GC-MS (70 eV), t_R = 18.158 min, m/z 265 (<1%); 250 [M-15]$^+$ (2%); 224 [M-41]$^+$ (50%); 73 [M-192]$^+$ (100%).

1-Nitro-4-[1-[(trimethylsilyl)oxy]-3-buten-1-yl]benzene (4c): IR ($CDCl_3$) 3079, 2958, 2903, 1641, 1607, 1522, 1492, 1432, 1416, 1350, 1316, 1292, 1253 cm^{-1}; ^{1}H NMR (300 MHz, $CDCl_3$) δ 8.19 (d, J = 8.29 Hz, 2H), 7.49 (d, J = 8.29 Hz, 2H), 5.74 (m, 1H), 5.03 (m, 2H), 4.80 (dd, J = 5.65, 6.78 Hz, 1H), 2.45 (m, 2H),

0.08 (s, 9H); ^{13}C NMR (75 MHz, CDCl$_3$) δ 152.37, 147.06, 133.93, 126.63, 123.43, 117.85, 73.85, 44.79, 0.02; GC-MS (70 eV), t_R = 20.424 min, m/z 265 M$^+$ (< 1%); 224 [M-41]$^+$ (70%); 73 [M-192]$^+$ (100%).

1-Fluoro-3-[1-[(trimethylsilyl)oxy]-3-buten-1-yl]benzene (4d): IR (CDCl$_3$) 3078, 2957, 2903, 1704, 1642, 1616, 1593, 1489, 1450, 1359, 1252 cm^{-1}; ^{1}H NMR (300 MHz, CDCl$_3$) δ 7.30 (m, 1H), 7.10 (m, 2H), 6.95 (m, 1H), 5.80 (m, 1H), 5.06 (m, 2H), 4.72 (dd, J = 5.46, 7.16 Hz, 1H), 2.48 (m, 2H), 0.11 (s, 9H); ^{13}C NMR (75 MHz, CDCl$_3$) δ 147.73, 146.57, 134.77, 130.20, 121.4, 118.94, 113.97, 112.61, 72.56, 43.82, 0.06; GC-MS (70 eV), t_R = 8.711 min, m/z 238 M$^+$ (<1%); 223 [M-15]$^+$ (5%); 197 [M-41]$^+$ (100%); 73 [M-165]$^+$ (95%).

1-Chloro-4-[1-[(trimethylsilyl)oxy]-3-buten-1-yl]benzene (4e): IR (CDCl$_3$) 3078, 2958, 2903, 1641, 1598, 1491, 1432, 1409, 1360, 1295, 1262, 1252 cm^{-1}; ^{1}H NMR (300 MHz, CDCl$_3$) δ 7.30 (m, 4H), 5.77 (m, 1H), 5.05 (m, 2H), 4.68 (dd, J = 5.46, 7.35 Hz, 1H), 2.43 (m, 2H), 0.07 s, 9H); ^{13}C NMR (75 MHz, CDCl$_3$) δ 143.41, 134.77, 132.62, 128.06, 127.28, 117.20, 74.17, 45.04, 0.10; GC-MS (70 eV), t_R = 11.263 min, m/z 254/256 M$^+$/M+2 (<1%); 213/215 [M-41]$^+$ (98%); 73 [M-181]$^+$ (100%).

1,3-Dichloro-2-[1-[(trimethylsilyl)oxy]-3-buten-1-yl]benzene (4f): IR (CDCl$_3$) 3078, 2958, 1642, 1581, 1562, 1437, 1251 cm^{-1}; ^{1}H NMR (300 MHz, CDCl$_3$) δ 7.29 (m, 2H), 7.11 (m, 1H), 5.81 (m, 1H), 5.50 (m, 1H), 5.05 (dd, J = 6.40, 8.10 Hz, 1H), 2.86 (m, 1H), 2.68 (m, 1H), 0.04 (s, 9H); ^{13}C NMR (75 MHz, CDCl$_3$) δ 137.92, 134.73, 129.38, 128.93, 128.60, 117.16, 71.45, 40.00, 0.12; GC-MS (70 eV), t_R = 13.209 min, m/z 288/290/292 M$^+$/M+2/M+4 (<1%); 247/249/251 [M-41]$^+$ (93%); 73 [M-215]$^+$ (100%).

3-[1-[(Trimethylsilyl)oxy]-3-buten-1-yl]benzonitrile (4g): IR (CDCl$_3$) 3078, 2958, 2902, 2230, 1641, 1602, 1584, 1482, 1434, 1359, 1285, 1252 cm^{-1}; ^{1}H NMR (300 MHz, CDCl$_3$) δ 7.62 (m, 1H), 7.53 (m, 2H), 7.42 (m, 1H), 5.73 (m, 1H), 5.01 (m, 2H), 4.72 (dd, J = 5.65, 6.78 Hz, 1H), 2.41 (m, 2H), 0.06 (s, 9H); ^{13}C NMR (75 MHz, CDCl$_3$) δ 146.41, 134.01, 130.76, 130.33, 129.55, 119.00, 117.79, 112.16, 73.73, 44.89, 0.04; GC-MS (70 eV), t_R = 14.267, m/z 245 M$^+$ (<1%); 230 [M-15]$^+$ (10%); 204 [M-41]$^+$ (98%); 73 [M-172]$^+$ (100%).

REFERENCES

1. Yamamoto, Y., Asso, N., *Chem Rev.* 1993, 93, 2207.
2. (a) Judd, W. R., Ban, S., Aube, J., *J. Am. Chem. Soc.* 2006, 128, 13736; (b) Pospisil, J., Kumamoto, T., Marko, I. E., *Angew. Chem. Int. Ed.* 2006, 45, 3357; (c) Wadamoto, M., Yamamoto, H., *J. Am. Chem. Soc.* 2005, 127, 14556; (d) Dong, W., *Youji Hauxue* 2001, 21, 1090; (e) Bonini, B. F., Comes-Franchini, M., Fochi, M., Mazzanti, G., Ricci, A., Varchi, G., *Tetrahedron Asymm.* 1998, 9, 2979; (f) Shing, T. K. M., Li, L.-H., *J. Org. Chem.* 1997, 62, 1230; (g) Matsumoto, K., Oshima, K., Utimoto, K., *J. Org. Chem.* 1994, 59, 7152; (h) Hayashi, T., Kabeta, K., Hamachi, I., Kumada, M., *Tetrahedron Lett.* 1983, 24, 2865.
3. (a) Hosomi, A., Sakurai, H., *Tetrahedron Lett.* 1976, 16, 1295; (b) Denmark, S. E., Almstead, N. G., in *Modern Carbonyl Chemistry*, ed. J. Otera, Wiley-VCH, Weinheim, Germany, 2000, chap. 11, p. 403.

4. (a) Kawabata, T., Kato, M., Mizugaki, T., Ebitani, K., Kaneda, K., *Chem. Eur. J.* 2005, 11, 288; (b) Jin, Y. Z., Yasuda, N., Furuno, H., Inanaga, J., *Tetrahedron Lett.* 2003, 44, 8765; (c) Sasidharan, M., Tatsumi, T., *Chem. Lett.* 2003, 32, 624.
5. Dintzner, M. R., Mondjinou, Y. A., Unger, B., *Tetrahedron Lett.* 2009, 50, 6639.
6. Nagendrappa, G., *Resonance* 2002, 64.
7. Dasgupta, S., Torok, B., *Org. Prep. Proc. Int.* 2008, 40, 1.
8. (a) Dintzner, M. R., Little, A. J., Pacilli, M., Pileggi D., Osner, Z. R., Lyons, T. W., *Tetrahedron Lett.* 2007, 48, 1577; (b) Dintzner, M. R., Wucka, P., Lyons, T. W., *J. Chem. Educ.* 2006, 83, 270; (c) Dintzner, M. R., Lyons, T. W., Akroush, M. H., Wucka, P., Rzepka, A. T., *Synlett* 2005, 5, 785; (d) Dintzner, M. R., McClelland, K. M., Morse, K. M., Akroush, M. H., *Synlett* 2004, 11, 2028; (e) Dintzner, M. R., Morse, K. M., McClelland, K. M., Coligado, D. M., *Tetrahedron Lett.* 2004, 45, 79.
9. Yahiaoui, A., Belbachir, M., Hachemaoui, A., *Int. J. Mol. Sci.* 2003, 4, 458.
10. For selected examples of the Hosomi–Sakurai reaction used in natural products synthesis see: (a) Troast, D. M., Yuan, J., Porco Jr., J. A., *Adv. Synth. Catal.* 2008, 350, 1701; (b) Spangenber, T., Airiau, E., Thuong, M. B. T., Donnard, M., Billet, M., Mann, A., *Synlett* 2008, 18, 2859; (c) Kalidindi, S., Jeong, W. B., Schall, A., Bandichhor, R., Nosse, B., Reiser, O., *Natl. Prod.* 2007, 46, 6361; (d) Pospisil, J., Marko, I. E., *J. Am. Chem. Soc.* 2007, 129, 3516; (e) van Innis, L., Plancher, J. M., Marko, I. E., *Org. Lett.* 2006, 8, 6111; (f) Crimmins, M. T., Zuccarello, J. L., Cleary, P. A., Parrish, J. D., *Org. Lett.* 2005, 8, 159; (g) Wender, P. A., Hegde, S. G., Hubbard, R. D., Zhang, L., *J. Am. Chem. Soc.* 2002, 124, 4956; (h) Williams, D. R., Myers, B. J., Mi, L., *Org. Lett.* 2000, 7, 945; (i) Bonjoch, J., Cuesta, J., Diaz, S., Gonzalez, A., *Tetrahedron Lett.* 2000, 41, 5669; (j) Davis, C. E., Duffy, B. C., Coates, R. M., *Org. Lett.* 2000, 2, 2717; (k) Marko, I. E., Mekhalfia, A., Murphy, F., Bayston, D., Bailey, M., Janousek, Z., Dolan, S., *Pure Appl. Chem.* 1997, 69, 565.
11. Fang, X., Watkin, J. G., Warner, B., *Tetrahedron Lett.* 2000, 41, 447.
12. For compound 4c see: (a) Ito, S., Hayashi, A., Komai, H., Yamaguchi, H., Kubota, Y., Asami, M., *Tetrahedron* 2011, 67, 2081; (b) Ito, S., Yamaguchi, H., Kubota, Y., Asami, M., *Tetrahedron Lett.* 2009, 50, 2967; (c) reference 5.
13. For compound 4h see: (a) reference 12a; (b) Yasuda, M., Fujibayashi, T., Shibata, I., Baba, A., Matsuda, H., Sonoda, N., *Chem. Lett.* 1995, 167; (c) Miyai, T., Inoue, K., Yasuda, M., Baba, A., *Synlett* 1997, 699; (d) Whitesell, J.K., Apodaca, R., *Tetrahedron Lett.* 1996, 37, 3955; (e) reference 11; (f) reference 12b; (g) reference 5.
14. Andraos, J., *Org. Proc. Res. Dev.* 2009, 13, 161.
15. Jiménez-González, C., Ponder, C. S., Broxterman, Q. B., Manley, J. B., *Org. Proc. Res. Dev.* 2011, 15, 912.

15 Environmentally Friendly Synthesis of Norstatines in Water

Ermal Ismalaj, Ferdinando Pizzo, and Luigi Vaccaro

CONTENTS

INTRODUCTION

α-Hydroxy-β-amino acids, **1**, are one of the most important moieties present in a wide number of biologically active compounds,[1] including both naturally occurring and synthetic pharmaceuticals. Among the numerous examples can be mentioned the aminopeptidase inhibitors (bestatin[2] and amastatin[3]), renin inhibitors (KRI-1314),[4] angiotensin-converting enzymes (microginin),[5] and antitumor agents (paclitaxel).[6]

Due to their importance, several procedures have been developed to synthesize α-hydroxy-β-amino acids and their derivatives, in both racemic and optically active form. These methods include asymmetric catalytic synthesis, enzymatic kinetic resolution, and the use of chiral auxiliaries and chiral building blocks.[1,7] The enantioselective synthesis α-hydroxy-β-amino acids have attracted a considerable amount of interest in recent years, and several synthetic approaches have been developed for the synthesis of these molecules.[7,8–16]

The aim of our work is to develop novel synthetic tools defining eco-sustainable reaction protocols by combining the use of green reaction media such as water[17] or solvent-free conditions[18] (SolFCs) and recoverable and reusable catalysts in order to reduce waste.

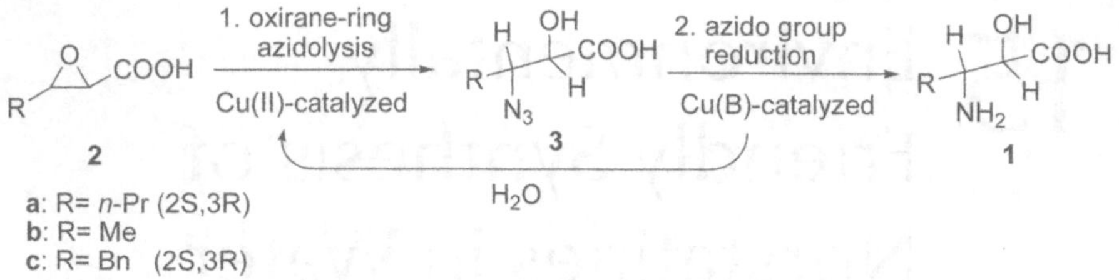

a: R= *n*-Pr (2S,3R)
b: R= Me
c: R= Bn (2S,3R)

SCHEME 15.1   One-pot Cu-catalyzed synthesis of α-hydroxy-β-amino acids **1** in water.

E-factor is one of the simplest metrics to measure waste production and has been widely adopted by the pharmaceutical industries[19] where development of greener synthetic routes for the preparation of biologically active molecules is highly appreciated.

We have defined an optimized green *one-pot* protocol for the synthesis of α-hydroxy-β-amino acids **1** based on the azidolysis of α,β-epoxycarboxylic acids **2** and the in situ reduction of the corresponding β-azido-α-hydroxy carboxylic acids **3**. Chemical efficiency and minimal environmental impact have been achieved with the definition of a specific synthetic strategy based on the use of one copper catalyst for both processes. In fact, the optimal efficiency of the azidolysis step in water has been reached with the use of a copper(II) salt (Cu(NO$_3$)$_2$), 10 mol%), which under basic conditions in the presence of sodium borohydride forms the corresponding solid copper boride Cu(B) that promotes the subsequent azido group reduction step (Scheme 15.1). The procedure for the recovery and reuse of the catalyst has been defined without any loss of its activity.

REACTION PROCEDURES

REPRESENTATIVE PROCEDURE FOR THE ONE-POT COPPER-CATALYZED SYNTHESIS OF α-HYDROXY-β-AMINO ACIDS BY A REACTION IN WATER

(+)-(2S,3R)-2,3-Epoxyhexanoic acid (**2a**) (note 1, 50 mmol, 6.50 g) was dissolved in water (50 mL), and powdered NaN$_3$ (note 2, 75 mmol, 4.88 g) was added under stirring. Then Cu(NO$_3$)$_2$ (note 3, 5 mmol, 0.94 g) was added as an aqueous solution (50 mL, 0.1 M) and the resulting pH of the aqueous mixture ranged between 4.3 and 4.5. The reaction mixture was warmed to 65°C, and after 1.5 h, the reaction mixture was cooled to 0°C in an ice bath and NaBH$_4$ (note 4, 100 mmol, 3.78 g) was added portion-wise. After 30 min at 0°C the reaction mixture was filtered and the copper boride was separated quantitatively (0.38 g). Some water (12.5 mL) was used to wash the solid catalyst and fully recover the reaction mixture. The aqueous layer was neutralized with HCl (note 5, 7 mL), and excess of NaBH$_4$ was consumed and some water was recovered by distillation (102–104 mL) to be reused in consecutive runs. The concentrated aqueous solution was acidified to pH 2 (note 6) using HCl (note 5, 5 mL) and subjected to ion-exchange column chromatography (note 7, resin Dowex 50W8X-400, 58.5 g). Elution was performed with NH$_4$OH 1M (note 8, 100 mL), and (–)-(2S,3S)-3-amino-2-hydroxyhexanoic acid (**1a**) was obtained in 83% yield (6.10 g) (note 9) as a white crystalline solid after water evaporation.

SECOND CYCLE USING RECOVERED CU(B) AND WATER

One hundred milliliters of the water recovered (see above) was taken and the pH was adjusted to 2 by adding a HCl (0.03 mL). Then, the recovered solid Cu(B) (0.38 g) was dispersed into the acidic recovered water. The mixture was left under magnetic stirring at 30°C for 2 h in open air, until complete dissolution of Cu(B) was reached. The process can be monitored by following the mixture color: at the beginning the mixture is brown, and after approximately 0.5 h it turns into a greenish color corresponding to the formation of (Cu²⁺). After 2 h, the mixture turns blue, confirming the complete formation of oxidized Cu^{2+} ions. At this point (+)-(2S,3R)-2,3-epoxyhexanoic acid (**2a**) (50 mmol, 6.5 g) and NaN_3 (75 mmol, 4.88 g) were added under stirring and the resulting pH was ca. 4.5. The reaction mixture was warmed to 65°C, and after 1.5 h the reaction mixture was cooled to 0°C (resulting pH 5.3–5.5). $NaBH_4$ (100 mmol, 3.78 g) was added portion-wise. After 30 min the reaction mixture was filtered and the Cu(B) was completely recovered. Some water (12.5 mL) was used to wash the catalyst and fully recover the reaction mixture. The aqueous layer was neutralized with HCl (7 mL), and the excess of $NaBH_4$ was consumed and some water (102–104 mL) was recovered by distillation. The concentrated aqueous solution was acidified at pH 2 using HCl (5 mL) and subjected to ion-exchange chromatography (Dowex resin 50W8X-400, 58.5 g). Elution was performed with 1 M NH_4OH (100 mL), and (–)-(2S,3S)-3-amino-2-hydroxyhexanoic acid (**1a**) was obtained in 83% yield (6.10 g) as a white crystalline solid after water evaporation.

NOTES

1. (+)-(2S,3R)-2,3-Epoxyhexanoic acid (**2a**) was prepared by Sharpless asymmetric epoxidation[8b] of the corresponding (E)-hex-2-en-1-ol-hexen-1-ol (97%), which was purchased from Sigma Aldrich, and subsequent Ru-catalyzed oxidation of the resulting epoxy alcohal **5a**.[8a] The protocols were performed on 50 mmol batches.
   Synthetic procedure for (+)-(2S,3R)-2,3-epoxyhexanoic acid (**2a**):
   Sharpless asymmetric epoxidation:

In an over-dried three-necked round-bottomed flask 175 mL of dry CH_2Cl_2 was added. After cooling the flask at –20°C (–)(S,S) diethyl tartrate (DET) (3 mmol, 0.62 g) and Ti(O-iPr)₄ (2.5 mmol, 0.71 g) were added under stirring and under nitrogen. t-Butyl hydroperoxide (note 10, 100 mmol, 19.5 mL) was added through the addition funnel. (E)-hex-2-en-1-ol (**4a**) (50 mmol, 5 g) was added and the mixture continued stirring for 2.5 h.

Workup: A freshly prepared solution of ferrous sulfate heptahydrate (60 mmol, 16.5 g) and tartaric acid (0.03 mmol, 5 g) in 100 mL of deionized water were cooled to 0°C. When the epoxidation reaction mixture reached 0°C, it was poured into a beaker containing the ferrous sulfate solution. The mixture was stirred for 10 min and then transferred to a separatory funnel. After phase separation the aqueous layer was extracted with 2 × 15 mL portions of ether. The combined organic layers were treated with 5 mL of a solution of NaOH at 30% (w/v) in saturated brine and stirred vigorously for 1 h. After transferring to a separatory funnel and diluting with 25 mL of water, phase separation occurred and the aqueous layer was extracted with 2 × 25 mL of ether. The combined organic layers were dried over sodium sulfate (0.340 g), filtered, and concentrated. The crude product was distilled through a 10 cm Vigreux column (the receiving flask was cooled in an ice bath), and (2R,3R)-3-propyloxiranemethanol **5a** was obtained in 85% yield and 93% ee.

Sharpless Ru-catalyzed oxidation of epoxyalcohol **5a** to epoxyacid **2a**:

5a   **2a**

This protocol was performed on a 50 mmol batch. (2R,3R)-3-propyloxiranemethanol **5a** (50 mmol, 5.8 g) was dissolved in CCl_4 (100 mL), CH_3CN (100 mL), and water (150 mL). Sodium metaperiodate (150 mmol, 32.08 g) was added under stirring. To the resulting biphasic solution ruthenium trichloride trihydrate (2.2 mol%, 0.287 g) was added, and the reaction mixture was stirred for 1.5 h at room temperature. 500 mL of dichloromethane was added and the phases were separated. The upper phase was extracted with 3 × 50 mL dichloromethane, the combined organic layers were dried over Na_2SO_4 (0.3 g) and passed through Celite (1.45 g), and removal of the solvent under vacuum gave **2a** in 92% yield.

2. Sodium azide (99%) was purchased from Sigma Aldrich and used without further purification.
3. Cupric nitrate (99% to 104%) was purchased from Sigma Aldrich and used without further purification.
4. Sodium borohydride (≥98%) was purchased from Sigma Aldrich and used without further purification.

5. Hydrochloric acid (38%) was purchased from Panreac Quimica and used without further purification.
6. The pH was measured using a Denver Instrument Ultra Basic-10 pH/mv meter.
7. Dowex Resin 50W8X-200-400 Hydrogen Form was obtained from Sigma Aldrich and used after treatment with 10 mL of concentrated HCl to adjust the pH.
8. Ammonium hydroxide (1 M) was prepared using 88 mL of water and 12 mL of a 9 M solution of NH_4OH, which was purchased from Sigma Aldrich.
9. The progress of the reaction was controlled by thin-layer chromatography (TLC) using silica on TLC Alu foils purchased from Sigma Aldrich. A mixture of MeOH/AcOH/H_2O (4/1/1) was used as an eluent and ninhydrin was used as a revealing agent.
10. *tert*-Butyl hydroperoxide solution in water (5.0–6.0 M) was purchased from Sigma Aldrich and used without further purification. By using the iodine liberation method,[27] the exact concentration of the epoxide was found to be 5.13 M. Detailed NMR data information:

> (−)-(2S,3S)-3-amino-2-hydroxyhexanoic acid (**1a**) (83% yield): ^{1}H NMR (400 MHz, D_2O) δ 0.85 (t, 3H, J = 7.5 Hz), 1.15–1.50 (m, 4H), 3.51 (m, 1H), 4.14 (d, 1H, J = 3.4 Hz); ^{13}C NMR (100.6 MHz, D_2O) δ 12.8, 18.1, 26.2, 53.6, 71.6, 176.6.

DISCUSSION

The one-pot copper-catalyzed synthesis of α-hydroxy-β-amino acids **1a–c** was achieved by azidolysis of α,β-epoxycarboxylic acids **2a–c** and subsequent reduction of the azido group of the corresponding α-hydroxy-β-azido acids **3a–c**. The results obtained are reported in Table 15.1.

The general one-pot procedure worked as follows. Initially, the azidolysis of α,β-epoxycarboxylic acids **2a–c** was performed in water by using 10 mol% of Cu(NO$_3$)$_2$ as catalyst and in the presence of 1.5 mol/equivalent of NaN$_3$. As the ring-opening process proceeds, the pH of the aqueous medium slightly increases (alkoxide formation) from ca. 4.5 to ca. 5.5, and at the end of this step (0.25–2 h, 65°C) 2.0 mol/equivalent of NaBH$_4$ was added at 0°C with the immediate increase of the pH over 10 and the formation of Cu(B) as a solid black precipitate. Subsequent reduction of the azido group proceeded smoothly in 0.5 h at 0°C. In the case of **2a**, the azidolysis was complete in 1.5 h, as described above, and the desired product **1a** was obtained in 83% yield, after its isolation in a pure form by a simple ion-exchange resin purification and avoiding the use of any organic solvent (Table 15.1, entry 1). **2b** and **2c** gave, using the same quantities of Cu(NO$_3$)$_3$, NaN$_3$, and NaBH$_4$, the expected α-hydroxy-β-amino acids **1b** and **1c** in 80% and 75% yields with the complete preservation of the purity of the chiral centers (Table 15.2, entries 2 and 3). The entire Cu-catalytic cycle can be realized only if water is used as reaction medium, and its use is essential in the recovery of the catalyst and its reconversion into Cu^{2+} in acidic solution. In fact, the solid black Cu(B) catalyst can be oxidized into Cu^{2+} if dissolved in an acidic aqueous solution (prepared using the recovered water) under air atmosphere (Scheme 15.2).[8a]

TABLE 15.1
Synthesis of α,β-Amino Acids 1a–c Using 10 mol% of Cu(NO$_3$)$_2$ as a Catalyst in Water

		t (h)			
Entry	Epoxide	Azidolysis	Reduction	Product	Yield (%)[a]
1	2a	1.5	0.5	1a	83
2	2b	2.0	0.5	1b	80
3	2c	0.25	0.5	1c	75

[a] Isolated yield of the pure products 1a–c.

TABLE 15.2
Summary of Metrics for One-Pot Reaction according to Scheme 15.3 and Experimental Protocols

Plan	% AE	% Yield	E-Kernel	E-Excess	E-Aux	E-Total	PMI
First cycle	43.1	83	1.8	8.4	37.9	48.1	49.1
Second cycle	43.1	83	1.8	16.6	29.6	48.0	49.0

This allows the reuse of the catalytic system in consecutive runs by simply adding new epoxide **2** and NaN$_3$ to repeat the Cu cycle (Scheme 15.2). Both the reaction medium and the catalyst were reused for five cycles without loss of efficiency and were always completely recovered.

IMPORTANCE ON GREEN CHEMISTRY CONTEXT

The efficiency of our synthetic route to optically active α-hydroxy-β-amino acids **1a–c** was evaluated and compared to other available synthetic protocols, by calculating the most relevant green metrics performances and according to the previously described Andraos algorithm, which has been used to test the greenness of several industrial and academic synthetic plans to different target molecules.[20] The examined metrics are percent overall yield, percent overall atom economy (% AE),[21] reaction

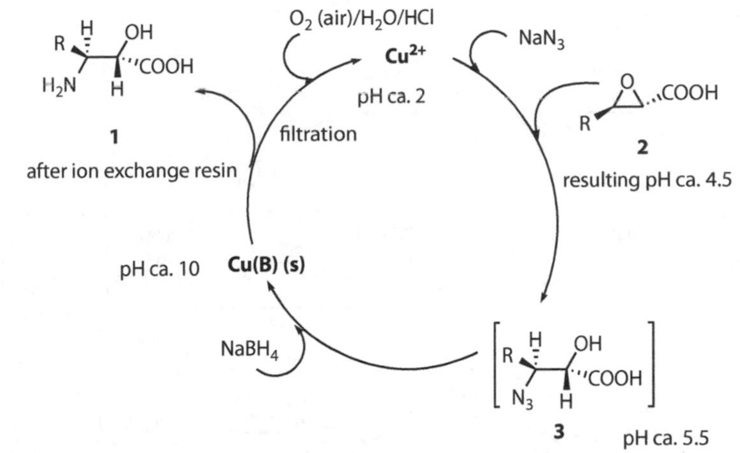

SCHEME 15.2 Copper catalytic cycle.

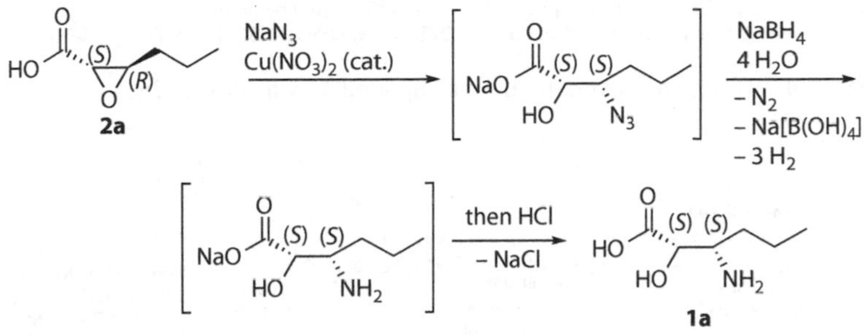

SCHEME 15.3 Balanced chemical equation for one-pot reaction used to calculate green metrics.

mass efficiency (RME),[22] overall E-factor,[23] and process mass intensity (PMI).[24] The overall E-factor in turn was subdivided into its components arising from by-products, side products, and unreacted starting materials (E-kernel), excess reagent consumption (E-excess), and auxiliary material consumption arising from reaction solvent, catalysts, all workup materials, and all purification materials (E-aux). Material efficiency metrics were carried out for the following: (1) the one-pot protocol involving epoxide ring opening of epoxide **2a** with azide, followed by reduction of the azido hydroxy acid **3a** intermediate by sodium borohydride to produce amino hydroxy acid **1a** (see Scheme 15.3); (2) various literature routes to compound **1a** via epoxide **2a** (see Scheme 15.4); and (3) various literature routes to compound **1a** not passing through epoxide **2a** (see Scheme 15.5).

Table 15.2 summarizes the comparison of material efficiency performances of the one-pot reaction using virgin copper nitrate as catalyst for the first cycle and using recovered copper boride from the first cycle, which is subsequently converted to Cu^{2+} via oxidation under acidic conditions in the second cycle. These data were determined using the balanced chemical equation shown in Scheme 15.3. Both cycles

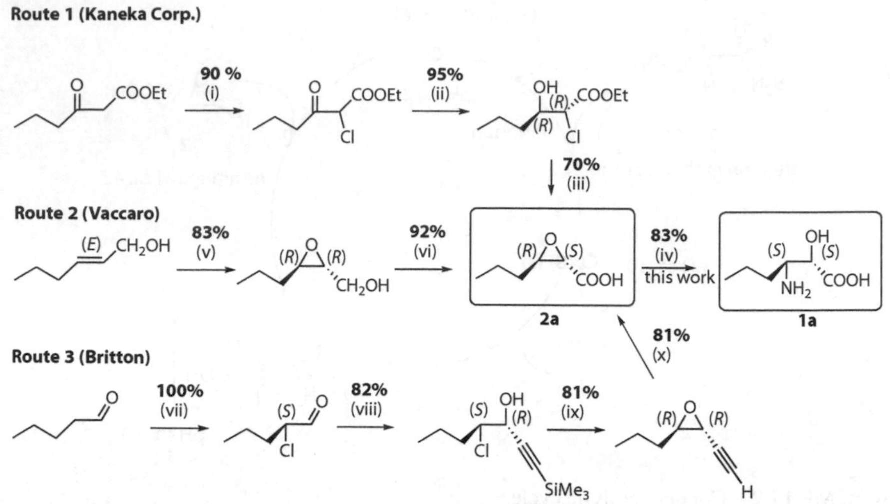

Route 1 (Kaneka Corp.)

Route 2 (Vaccaro)

Route 3 (Britton)

(i) SO$_2$Cl$_2$ (ii) E.coli HB101/FERM BP-10024 (iii) NaOEt/KOH then HCl (iv) Cu(NO$_3$)$_2$(cat.)/NaN$_3$, then NaBH$_4$ (v)tBuOOH/Ti(OiPr)$_4$(cat.)/(+)-DET (vi) NaIO$_4$/RuCl$_3$ 3 H$_2$O (vii) NBS/(R)-prolinamide (viii) ethynyltrimethylsilane/nBuLi, then NH$_4$Cl (ix) Cs$_2$CO$_3$/EtOH (x) KHSO$_5$ KHSO$_4$ K$_2$SO$_4$/RuCl$_3$ 3 H$_2$O(cat.)

SCHEME 15.4   Reaction network to make compound **1a** via epoxide **2a**.

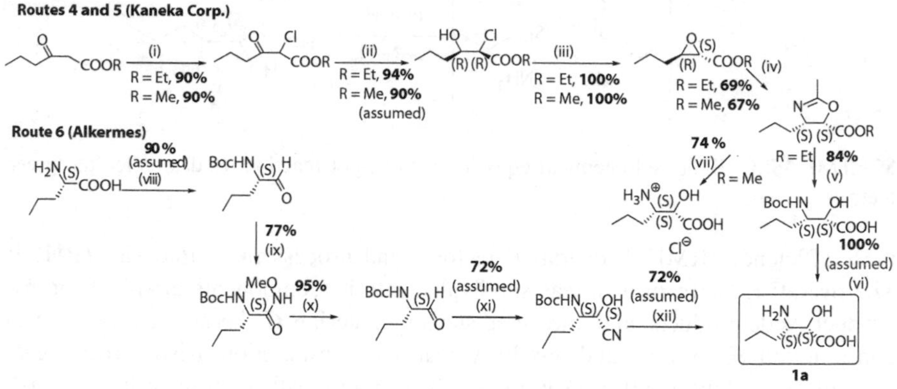

Routes 4 and 5 (Kaneka Corp.)

Route 6 (Alkermes)

(i) SO$_2$Cl$_2$ (ii) E. coli HB101/FERM BP-10024 (iii) K$_2$CO$_3$ (iv) CH$_3$CN/BF$_3$ OEt$_2$ (cat.) (v) HCl (aq), then Boc$_2$O (vi) HCl/H$_2$O-dioxane (vii) H$_2$O, HCl (viii) NaOH(cat.)/Boc$_2$O (ix) carbodiimidazole/MeONHCH$_3$ HCl/iPr$_2$NEt (x) LiAlH$_4$ (xi) NaHSO$_3$, then NaCN (xii) H$_2$O/HCl

SCHEME 15.5   Reaction network to make compound **1a** without going through epoxide **2a**.

result in essentially the same PMI value. Table 15.3 summarizes the metrics results for the reaction networks shown in Schemes 15.4 and 15.5. As shown in the table, our synthetic method to synthesize compound **1a** is the best among other literature reports to the same target product in terms of sustainability, giving a lower PMI, even including workup and purification. It should be pointed out that E-aux, E-total, and PMI values for literature plans listed in Table 15.3 were determined as lower limits since full disclosure of materials in workup and purification procedures was

(note 4, 20 mL) were heated at reflux while stirring continuously under nitrogen atmosphere. After 2 hours the reaction was stopped and cooled at room temperature. The reaction mixture was then filtrated over the glass filtration device Gooch no. 4 and the solvent evaporated, allowing the isolation of the pure product (0.83 g, 95% yield).

SYNTHESIS OF PHTHALAN

In a round-bottom flask equipped with a dephlegmator, 1.0 g dihydroxymethyl benzene (note 5, 7.24 mmol, 1 mol. equiv.), 2.6 g DMC (note 2, 29.0 mmol, 4 mol. equiv.), 0.76 g sodium methoxide (note 3, 14.4 mmol, 2 mol. equiv.), and acetonitrile (note 4, 20 mL) were heated at reflux while stirring continuously under nitrogen atmosphere. After 4 hours the reaction was stopped and cooled at room temperature. The reaction mixture was then filtrated over the glass filtration device Gooch no. 4 and the solvent evaporated, allowing the isolation of the pure product (0.83 g, 95% yield).

SYNTHESIS OF (–)-NORLABDANE OXIDE (AMBROXAN)

In a round-bottom flask equipped with a dephlegmator, 1.0 g amberlyn diol (note 6, 3.93 mmol, 1 mol. equiv.), 10.6 g DMC (note 2, 118.00 mmol, 30 mol. equiv.), and 0.88 g potassium *tert*-butoxide (note 7, 7.87 mmol, 2 mol. equiv.) were heated at reflux while stirring continuously under nitrogen atmosphere. The reaction was followed by thin-layer chromatography (TLC) until complete disappearance of the starting material. The solution was then filtered and DMC removed by evaporation to recover the pure ambroxan as a white crystalline powder (0.88 g, 95% yield).

SYNTHESIS OF ISOSORBIDE

In a round-bottom flask equipped with 2.0 g D-sorbitol (note 8, 10.98 mmol, 1 mol. equiv.), 8 g DMC (note 2, 87.88 mmol, 8 mol. equiv.), 2.4 g sodium methoxide (note 3, 43.42 mmol, 4 mol. equiv.), and methanol (note 9, 30 mL) were heated at reflux while stirring continuously under nitrogen atmosphere. After 8 hours the reaction was stopped, cooled at room temperature, and diethyl ether was added to the mixture. The reaction mixture was then filtrated over the glass filtration device Gooch no. 4 and the solvent evaporated to recover the pure isosorbide as light yellow crystals in 76% yield (1.22 g).

SYNTHESIS OF N-CARBOXYMETHYLINDOLINE (2,3-DIHYDRO-INDOLE-1-CARBOXYLIC ACID METHYL ESTER)

In a typical experiment, 2-(2-aminophenyl)ethanol (note 10, 0.5 mL, 3.64 mmol), DMC (note 2, 15 mL, 166.60 mmol), and potassium *tert*-butoxide (note 7, 2.5 mol. equiv.) were heated at T = 90°C while stirring continuously under nitrogen atmosphere for 3 hours. The reaction outcome was followed by gas chromatography–mass spectrometry (GC-MS) analysis. The reaction mixture was then filtered and the solvent evaporated under vacuum to recover the pure N-carboxymethylindoline (0.50 g, 78% yield).

SYNTHESIS OF N-CARBOXYMETHYLISOINDOLINE (1,3-DIHYDRO-ISOINDOLE-2-CARBOXYLIC ACID METHYL ESTER)

In a typical experiment, 2-(aminomethyl)benzyl alcohol (note 11, 0.5 mL, 3.64 mmol), DMC (note 2, 15 mL, 166.60 mmol), and potassium *tert*-butoxide (note 7, 2.5 mol. equiv.) were heated at T = 90°C while stirring continuously under nitrogen atmosphere for 6 hours. The reaction outcome was followed by GC-MS analysis. The reaction mixture was then filtered and the solvent evaporated under vacuum to recover the pure N-carboxymethylisoindoline (0.52 g, 80% yield).

NOTES

1. 2-Hydroxyethyl phenol (99%) was purchased from Sigma Aldrich and used without further purification.
2. DMC (99%) was purchased from Sigma Aldrich and used without further purification.
3. Sodium methoxide reagent grade (95%) powder was purchased from Sigma Aldrich and used without further purification.
4. Acetonitrile American Chemical Society (ACS) reagent (≥99.5%) was purchased from Sigma Aldrich and used without further purification.
5. Dihydroxymethyl benzene purum (≥95.0%) (high-performance liquid chromatography (HPLC)) was purchased from Fluka and used without further purification.
6. A pure sample of amberlyn diol was provided by Imperial Chemical Industries.
7. *tert*-Butoxide (reagent grade, ≥98%) was purchased from Sigma Aldrich and used without further purification.
8. D-Sorbitol (≥98%) was purchased from Sigma Aldrich and used without further purification.
9. Methanol ACS reagent (≥99.8%) was purchased from Sigma Aldrich and used without further purification.
10. 2-(2-Aminophenyl)ethanol (>97%) was purchased from Sigma Aldrich and used without further purification.
11. 2-(Aminomethyl)benzyl alcohol (>97%) was purchased from Sigma Aldrich and used without further purification.

DISCUSSION

A range of aliphatic and aromatic 1,4-diols, each bearing differently substituted alcohols (primary, secondary, tertiary, allyl, phenyl, and benzyl), have shown to undergo fast and high-yielding cyclization reactions by DMC chemistry.[13] In particular, aliphatic 1,4-diols formed tetrahydrofuran derivatives in high yield, although isolation of the products is somehow difficult due to their very low boiling points; thus, these results are not reported in this work.

On the other hand, 1,4-diols bearing aromatic moieties, i.e., 1,2-dihydroxymethyl benzene and 2-hydroxyethylphenol, lead to more stable products, and their reactivity with DMC is reported in Table 16.1 (Scheme 16.1).

TABLE 16.1

Cyclization of 1,4-Diols and 4-Amino-1-Butanol Derivative with DMC to Heterocyclic Compounds

Entry	Substrate (gr)	Solvent (ml)	DMC eq. mol	Base (eq. mol)	Time h	Heterocyclic Product yield % (Isolated)
1	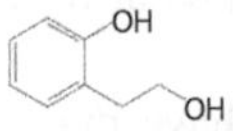 1	CH₃CN (20)	4	NaOMe (2)	2	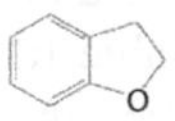 100 (95)
2	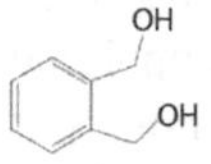 1	CH₃CN (20)	4	NaOMe (2)	4	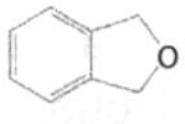 100 (95)
3	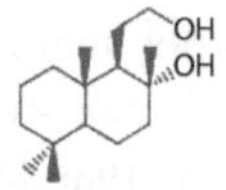 1	–	30	t-BuOK (2)	2	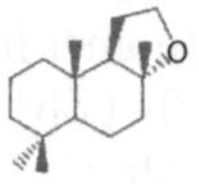 100 (95)
4	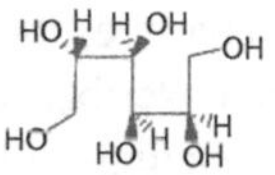 2	MeOH (30)	8	NaOMe (4)	8	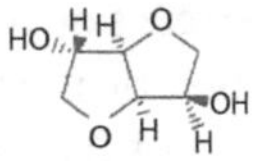 98 (76)
5	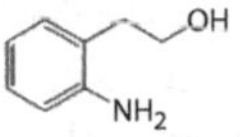 0.5	–	45	t-BuOK (2.5)	3	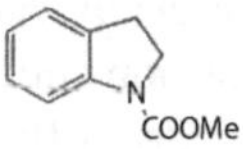 95 (78)
6	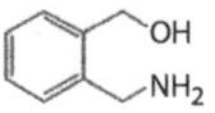 0.5	–	45	t-BuOK (2.5)	6	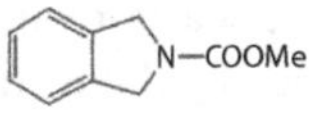 95 (80)

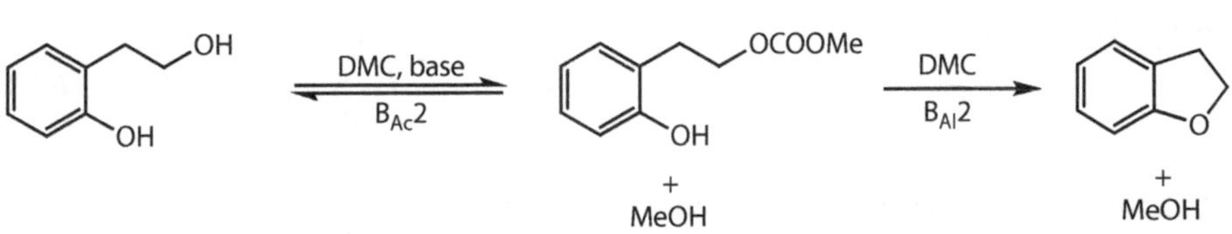

SCHEME 16.1 Cyclization of 2-hydroxyethyl phenol to 2,3-dihydrobenzofuran.

2-Hydroxyethyl phenol ended up being a very reactive substrate, and in the presence of a base and DMC, it forms 2,3-dihydrobenzofuran in quantitative yield (Table 16.1, entry 1).

1,2-Bis(hydroxymethyl)benzene also led to the quantitative formation of the related cyclic ether, phthalan, under reaction conditions similar to those used for 2-hydroxyethylphenol (Table 16.1, entry 2).

Norlabdane oxide, one of the preferred synthetic compounds with a desirable ambergris-type odor, is commercially available under various trade names. The industrial synthesis of norlabdane oxide by cyclization of the related diol, amberlyn diol, under acidic conditions leads to a mixture of ambroxan (ca. 60%) and by-products derived from the concurrent elimination reaction.[14]

However, ambroxan has been shown to be efficiently synthesized by reaction of amberlyn diol with DMC used as both a reagent and a solvent in the presence of a base (Table 16.1, entry 3).

Cyclic ethers in the form of anhydro sugar alcohols have many applications in industry, in particular in the food and pharmaceutical industries, and they are also employed as monomers in the formation of polymers and copolymers. Such anhydro sugar alcohols are derivatives of mannitol, iditol, and sorbitol.

In particular, isosorbide, the anhydro sugar alcohol derived from sorbitol, is useful as a monomer in the manufacture of many derivatives as polymers and copolymers, especially polyester polymers and copolymers. Isosorbide is industrially synthesized by dehydration of sorbitol by an acid-catalyzed reaction that leads to different anhydro compounds, but also to polymer-like products.[15]

The reaction of D-sorbitol with DMC in the presence of a base resulted in a low-yield formation of isosorbide (Table 16.1, entry 4). This has been ascribed to the chemical behavior of isosorbide; that is, once formed, it keeps on reacting with excess DMC, leading to the formation of its methoxycarbonyl and methyl derivatives (Scheme 16.2). To prevent unwanted consecutive reactions, methanol was used as a solvent. In this case, the reaction equilibrium is shifted toward isosorbide, avoiding the formation of its derivatives. Under these conditions, isosorbide formed readily and in high yield (Table 16.1).

The mechanism of the 1,4-diol cyclization follows in any case a two-step reaction: methylcarbonylation of the starting diol, followed by intramolecular nucleophilic attack by the tertiary alcohol.

Under the studied reaction conditions two simple aromatic bifunctional nucleophiles (4-amino-1-alcohols), i.e., 2-(2-aminophenyl)ethanol and 2-(aminomethyl) benzyl alcohol, were also investigated (Scheme 16.3). Results, reported in Table 16.1,

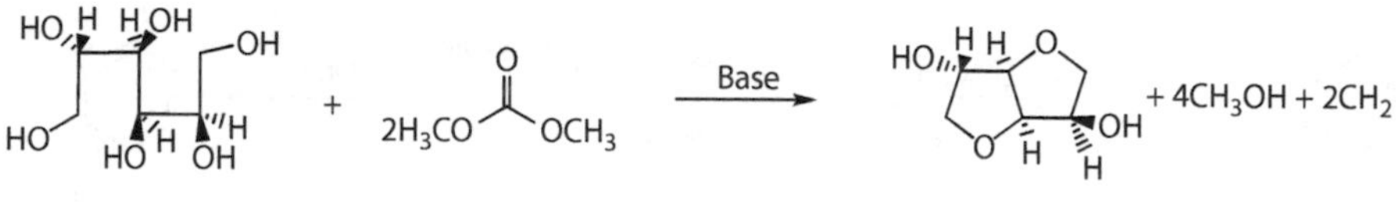

SCHEME 16.2   Cyclization of D-sorbitol to isosorbide.

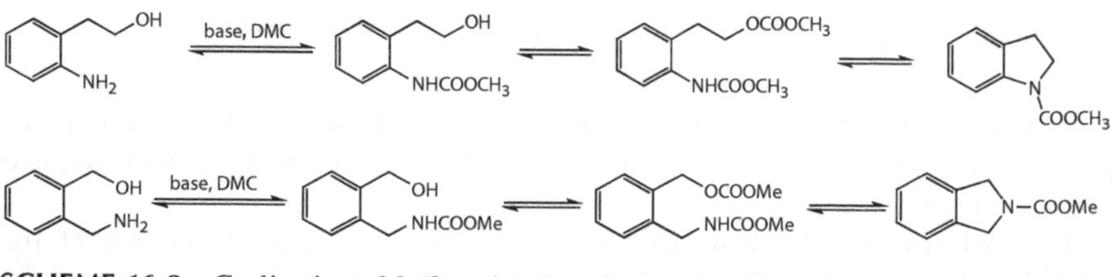

SCHEME 16.3   Cyclization of 2-(2-aminophenyl)ethanol to N-carboxymethylindoline and 2-(aminomethyl)benzyl alcohol to N-carboxymethylisoindoline.

demonstrated that also in this case N-carboxymethyl indoline (Table 16.1, entry 5) and N-carboxymethylisoindoline (Table 16.1, entry 6) were formed in quantitative yield by intramolecular cyclization.[16]

In this case the 4-amino-1-butanol derivative, most probably, first undergoes carboxymethylation at the hydroxyl (or amine) group ($B_{Ac}2$), then, or at the same time, the amine (or hydroxyl) moiety also carboxymethylates ($B_{Ac}2$). As a consequence, the amino group of the so-formed carbamate results in a softer nucleophilic character, and it undergoes fast alkylation to form selectively the carboxymethyl pyrrolidine ($B_{Al}2$). The formation of the N-based heterocycle is favored because all the carboxy-methylation reactions ($B_{Ac}2$) are in equilibria, except for the one related to the cyclic formation ($B_{Al}2$).

GREEN METRICS

Since there were no competing synthesis plans in the literature using the described cyclization strategy to make N-carboxymethylindoline and N-carboxymethylisoindoline, green metrics analyses were performed on various synthesis plans to make 2,3-dihydrobenzofuran, phthalan, ambroxan, and isosorbide. The results for the four compounds are summarized in Tables 16.2 to 16.5. In each table plans are ranked in ascending order of process mass intensity (PMI).[17] Entries denoted with minimum estimates for PMI indicate that not all materials used in the experimental procedures for the purification phases of the reactions were disclosed. These deficiencies impacted reliable estimates of E-aux and E-total.[18]

With respect to the Glaxo procedure to make 2,3-dihydrobenzofuran using the Vilsmeier–Haack reagent (chloromethylene-dimethyl-ammonium chloride), no experimental procedure was given, and hence all E-factor parameters were not determinable (see Table 16.2, entry 7). The present strategy to make 2,3-dihydrobenzofuran ranks second in material efficiency behind solvent-free dehydration catalyzed by alumina on silica gel at elevated temperature (see Table 16.2, entries 1 and 2). The Mitsunobu strategy has the lowest atom economy (see Table 16.2, entry 3). From an energy consumption perspective, the present DMC method is run at temperatures that are significantly lower than prior reports, with the exception of the Mitsunobu method, which is run at room temperature. It appears that the DMC method is the best overall compromise satisfying both material and energy efficiencies for the synthesis of 2,3-dihydrobenzofuran. From Table 16.3, the present cyclization strategy leading to phthalan ranks first in material efficiency (see entry 1) with the highest yield, though its atom economy is significantly lower than that of the dehydration methodologies using a solid-supported catalyst at high temperature (see Table 16.3, entry 2) or an aqueous hydrochloric acid solution (see Table 16.3, entry 3). The sequential oxidation-reduction strategy using manganese dioxide and triethylsilane has the worst atom economy (see Table 16.3, entry 4). Ambroxan is most efficiently made by cyclizing ambradiol using DMC as a sacrificial reagent (see Table 16.4, entry 1). From Table 16.5, the first six entries correspond to minimum estimates of PMI, which makes the double cyclization of D-sorbitol (also known as D-glucitol in the literature) to isosorbide via DMC the most reliably determined efficient strategy (see entry 8) despite its

TABLE 16.2

Summary of Metrics for Various Syntheses of 2,3-Dihydrobenzofuran by Cyclizing 2-Hydroxyethyl Phenol

Plan	Conditions	% AE[a]	% Yield	E-Kernel	E-Excess	E-Aux	E-Total	PMI[b]
Method C (1980)[d]	250°C Alumina on silica gel	87.0	77	0.5	0	6.1	6.6	7.6
Tundo (this work)	DMC/NaOMe, 90°C	52.6	95	1.0	2.4	19.8	23.1	24.1
Kuznetsov (2012)[e]	PPh$_3$ EtOOC-N = N-COOEt, room temperature	20.9	80	5.0	1.4	>45.5	>51.9	>52.9
Method A (1980)[d]	NaNO$_2$/H$_2$SO$_4$, then NaOH, 50°C	27.1	36	9.2	0.9	48.4	58.6	59.6
Nagashima (2011)[f]	350°C [(Nb$_6$Cl$_{12}$)Cl$_2$(H$_2$O)$_4$] H$_2$O on silica	87.0	7	15.1	0	161.3	176.4	177.4
Method B (1980)[d]	250°C Silica gel	87.0	2	65	0	288	353	354
Glaxo (1993)[g]	Vilsmeier–Haack reagent/Et$_3$N, reaction temperature not stated	25.6	80	ND[c]	ND	ND	ND	ND

a AE = atom economy.

b PMI = process mass intensity.

c ND = not determinable

d Bakke, J.M., Roholdt, H.M., *Acta Chem. Scand. B*, 1980, 34, 73.

e Kuznetsov, A., Gevorgyan, V., *Org. Lett.*, 2012, 14, 914–917.

f Nagashima, S., Kamiguchi, S., Kudo, K., Sasaki, T., Chihara, T., *Chem. Lett.*, 2011, 40, 78–80.

g Procopiou, P.A., Brodie, A.C., Deal, M.J., Hayman, D.F., *Tetrahedron Lett.*, 1993, 34, 7483–7486.

TABLE 16.3
Summary of Metrics for Various Syntheses of Phthalan by Cyclizing Dihydroxymethyl Benzene

Plan	Conditions	% AE[a]	% Yield	E-Kernel	E-Excess	E-Aux	E-Total	PMI[b]
Tundo (this work)	DMC/NaOMe, 90°C	52.6	95	1.0	2.4	19.8	23.1	24.1
Nagashima (2011)[c]	350°C [(Nb$_6$Cl$_{12}$)Cl$_2$(H$_2$O)$_4$] H$_2$O on silica	87.0	24	3.8	0	47.9	51.7	52.7
Seebach (1980)[d]	HCl, room temperature	87.0	86	0.3	0	>265.2	>265.5	>266.5
Panda (2008)[e]	MnO$_2$, then Et$_3$SiH/CF$_3$COOH, −5°C to room temperature	35.2	91	2.1	6.9	>395.4	>404.5	>405.5

a AE = atom economy.
b PMI = process mass intensity.
c Nagashima, S., Kamiguchi, S., Kudo, K., Sasaki, T., Chihara, T., *Chem. Lett.*, 2011, 40, 78–80.
d Meyer, N., Seebach, D., *Chem. Ber.*, 1980, 113, 1304–1319.
e Panda, B., Sarkar, T.K., *Tetrahedron Lett.*, 2008, 49, 6701–6703.

TABLE 16.4
Summary of Metrics for Various One-Step Syntheses of Ambroxan by Cyclizing Ambradiol

Plan	Conditions	% AE[a]	% Yield	E-Kernel	E-Excess	E-Aux	E-Total	PMI[b]
Tundo (this work)	DMC	68.6	95	0.5	11.6	1.0	13.1	14.1
Moulines (2001)[d]	nBuLi TsCl	41.2	96	1.5	0.03	44.7	46.2	47.2
Barrero (1993)[e]	TsCl Pyridine	39.2	99	1.6	7.9	81.9	91.4	92.4
de Groot (1994)[f]	pTsOH	92.9	80	0.3	0	>267.5	>267.8	>268.8
Akita (2000)[g]	pTsOH H$_2$O	92.9	59	0.8	0	>288.5	>289.4	>290.4
Buchi (1989)[h]	pTsOH	92.9	75	0.4	0	>335.6	>336.0	>337.0
Akita (1998)[i]	pTsOH H$_2$O	92.9	45	1.4	0	>352.0	>353.4	>354.4
Kutney (1997)[j]	pTsOH	92.9	76	0.4	0	>468.6	>469.0	>470.0
de Groot (2001)[k]	pTsOH	92.9	86	0.3	0	>529.9	>530.2	>531.2
Barrero (2004)[l]	PPh$_3$ BrCCl$_3$ NaHCO$_3$	29.6	57	4.9	12.9	>722.2	>740.0	>741.0
Castro (2002)[m]	pTsOH H$_2$O	92.9	75	0.4	0	>801.3	>801.8	>802.8
Koga (1998)[n]	TsCl Pyridine	39.2	73	2.5	84.9	>953.3	>1040.7	>1041.7
Kutney (1994)[c,o]	pTsOH	92.9	48	1.2	0	>2510.3	>2511.6	>2512.6
Mori (1990)[p]	TsCl Pyridine	39.2	78	2.3	72.4	>2496.3	>2571.0	>2572.0

[a] AE = atom economy.

[b] PMI = process mass intensity.

[c] Starting material is 4-(2-hydroxy-ethyl)-3,4a,8,8-tetramethyl-decahydro-naphthalen-2-ol.

[d] Moulines, J., Lamidey, A.M., Desvergnes-Breuil, V., *Synth. Commun.*, 2001, 31, 749–758.

[e] Barrero, A.F., Alvarez-Manzaneda, E.J., Altarejos, J., Salido, S., Ramos, J.M., *Tetrahedron*, 1993, 49, 10405–10412.

[f] Verstegen-Haaksma, A.A., Swarts, H.J., Jansen, B.J.M., de Groot, A., *Tetrahedron*, 1994, 50, 10095–10106.

[g] Akita, H., Nozawa, M., Mitsuda, A., Ohsawa, H., *Tetrahedron Asymm.*, 2000, 11, 1375–1388.

[h] Buchi, G., Wuest, H., *Helv. Chim. Acta*, 1989, 72, 996–1000.

[i] Akita, H., Nozawa, M., Shimizu, H., *Tetrahedron Asymm.*, 1998, 9, 1789–1799.

[j] Kutney, J.P., Cirera, C., *Can. J. Chem.*, 1997, 75, 1136–1150.

[k] Bolster, M.G., Jansen, B.J.M., de Groot, A., *Tetrahedron*, 2001, 57, 5657–5662.

[l] Barrero, A.F., Alvarez-Manzaneda, E.J., Chahboun, R., Arteaga, A.F., *Synth. Commun.*, 2004, 34, 3631–3643.

[m] Castro, J.M., Salido, S., Altarejos, J., Nogueras, M., Sanchez, A., *Tetrahedron*, 2002, 58, 5941–5949.

[n] Koga, T., Aoki, Y., Hirose, T., Nohira, H., *Tetrahedron Asymm.*, 1998, 9, 3819–3823.

[o] Kutney, J.P., Chen, Y.H., *Can. J. Chem.*, 1994, 72, 1570–1581.

[p] Mori, K., Tamura, H., *Liebigs Ann. Chem.*, 1990, 361–368.

TABLE 16.5

Summary of Metrics for Various Syntheses of Isosorbide by Cyclizing D-Sorbitol

Plan	Conditions	No. Steps	% AE[a]	% Yield	E-Kernel	E-Excess	E-Aux	E-Total	PMI[b]
Holloday (2007)[c,e]	Amberlyst 36 Pd/C	1	80.2	73	0.7	0	0.4	1.1	>2.1
Dupont (2003)[c,f]	H_2SO_4	1	80.2	80	0.6	0	>2.0	>2.5	>3.5
Montgomery (1946)[c-g]	HCl	1	80.2	66	0.9	0	>8.2	>9.1	>10.1
Defaye (1990)[c,h]	HCOOH HF	1	80.2	70	0.8	0	>9.4	>9.4	>10.4
Bock (1981)[i]	170°C Amberlite IR-120	1	80.2	57	1.2	0	>8.7	>9.9	>10.9
Hockett (1946)[c,j]	H_2SO_4	1	80.2	82	0.5	0	>20.3	>20.9	>21.9
Liu–Luckett (2008)[k]	250°C Pentasil zeolite	1	80.2	43	2.0	0	15.5	17.5	18.5
Tundo (this work)[c]	DMC, 90°C	1	40.3	76	2.3	4.9	21.4	28.6	29.6
Haworth (1948)[d,l]	1a. H_2SO_4 1b. H_2/Raney Ni/HCl 2. Fractional distillation	2	40.1	33.5	6.4	26.1	34.2	66.8	67.8
Kurszewska (2002)[c,m]	290°C Molecular sieves 3 A	1	80.2	44	1.9	0	82.0	>83.8	>84.8

[a] AE = atom economy.

[b] PMI = process mass intensity.

[c] Starting material is D-sorbitol; an undisclosed amount of excess barium carbonate was used to neutralize sulfuric acid.

[d] Starting material is sucrose.

[e] Holloday, J.E., Hu, J., Zhang, X., Wang, Y., U.S. 20071173654, 2007.

[f] Bhatia, K.K., U.S. 2003229235, Dupont Dow Elastomers, 2002; Bhatia, K.K. WO 03089436 Dupont, 2003; Bhatia, K.K. WO 03089445 Dupont, 2003.

[g] Montgomery, R., Wiggins, L.F., *J. Chem. Soc.*, 1946, 390–393.

[h] Defaye, C., Gadelle, A., Pedersen, C., *Carbohydrate Res.*, 1990, 205, 191–202.

[i] Bock, K., Pedersen, C., Thorgersen, H., *Acta Chem. Scand.*, 1981, 35B, 441–449.

[j] Hockett, R.C., Fletcher, G., Jr., Sheffield, E.L., Goepp, R.M., Jr., *J. Am. Chem. Soc.*, 1946, 68, 927–930.

[k] Liu, A., Luckett, C., U.S. 2008249323, 2008.

[l] Haworth, W.N., Wiggins, L.F., GB 600870, 1948.

[m] Kurszewska, M., Skorupowa, E., Madaj, J., Konitz, A., Wojnowski, W., Wisniewski, A., *Carbohydrate Res.*, 2002, 337, 1261–1268.

atom economy being half that of a simple double dehydration. Entry 7 corresponds to a reaction performance based on HPLC product studies without isolation of products.

CONCLUSIONS

In this work we report on the synthesis of five-membered N- and O-heterocycles via DMC chemistry. In all the examples discussed above, DMC acts as a sacrificial molecule, as the so-formed cyclic products do not include DMC or its functional groups in their structures except for the N-based heterocycles, where the methoxy-carbonyl derivatives are readily formed.

In particular, 1,4-diols were efficiently and quantitatively converted into cyclic ethers in high yield and in short reaction times by reaction with DMC in the presence of a base. This synthetic procedure was employed for simple cyclic ethers, as well as for industrially relevant compounds such as (–)-norlabdane oxide and isosorbide.

Furthermore, reacting aromatic bifunctional compounds, i.e., 4-amino-1-butanol derivatives, with DMC in the presence of a base and under mild conditions also resulted in the formation of the corresponding N-based heterocycles in high yield and in short reaction times.

Comparing this reaction methodology with a chlorine-based procedure, the DMC-mediated pathway is quantitative, occurs in one step, does not require any chlorine-based chemical or strong acid, and does not produce any chlorinated waste material.

REFERENCES

1. (a) S.V. Bhat, B.S. Bajwa, H. Dornauer, N.J. de Souza, H.-W. Fehlhaber, *Tetrahedron Lett.* 1977, 18, 1669–1672; (b) S.V. Bhat, B.S. Bajwa, H. Dornauer, N.J. de Souza, *J. Chem. Soc. Perkin Trans. 1* 1982, 767–771; (c) S.V. Bhat, *Prog. Chem. Org. Nat. Prod.* 1993, 1; (d) J.W. Westley, *Polyethers Antibiotics: Naturally Occurring Acid Ionophores*, vols. 1–2, Marcel Dekker, New York, 1982; (e) M. Imoto, K. Umezawa, Y. Takahashi, H. Naganawa, Y. Iitaka, H. Nakamura, Y. Koizurni, Y. Sasaki, M. Hamada, T. Sawa, T. Takeuchi, *J. Nat. Prod.* 1990, 53, 825–829; (f) H. Odai, K. Shindo, A. Odagawa, J. Mochizuki, M. Hamada, T. Takeuchi, *J. Antibiot.* 1994, 47, 939–941; (g) N.O. Fuller, J.P. Morken, *Org. Lett.* 2005, 7, 4867–4869; (h) G. Wang, F.J. Burczynski, B.B. Hasinoff, K. Zhang, Q. Lu, J.E. Anderson, *Mol. Pharm.* 2009, 6, 895–904.
2. J.K. Rossiter, *Chem. Rev.* 1996, 96, 3201.
3. (a) E.K. Lee, Y.H. Baek, Patent EP1939190, 2008, to Hysung Corporation; (b) I.I. Geiman, L.F. Bulenkova, A.A. Lazdin'sh, A.K. Veinberg, V.A. Slavinskaya, A.A. Avots, *Chem. Heterocycl. Compd.* 1981, 4, 314–316; (c) T. Shibata, R. Fujiwara, Y. Ueno, *Synlett* 2005, 152–154; (d) G.V. Sharma, K.R. Kumar, P. Sreenivas, P.R. Krishna, M.S. Chorghade, *Tetrahedron Asymm.* 2002, 13, 687–690; (e) R.I. Khusnutdinov, N.A. Shchadneva, A.R. Baiguzina, Y. Lavrentieva, U.M. Dzhemilev, *Russ. Chem. Bull.* 2002, 51, 2074–2079; (f) B. Panda, T.K. Sarkar, *Tetrahedron Lett.* 2008, 49, 6701–6703; (g) H. Zhao, J.E. Holladay, H. Brown, Z.C. Zhang, *Science* 2007, 316, 1597–1600.
4. O. Lindner, L. Rodefeld, *Benzenesulfonic Acids and Their Derivatives, Ullmann's Encyclopedia of Industrial Chemistry*, Wiley-VCH, Weinheim, 2001; (b) C.S. Salteris, I.D. Kostas, M. Micha-Screttas, G.A. Heropoulos, C.G. Screttas, A. Terzis, *J. Org. Chem.*

1999, 64, 5589–5592; (c) F. Ojima, T. Matsue, T. Osa, *Chem. Lett.* 1987, 2235–2238; (d) C. Ferreri, C. Costantino, C. Chatgilialoglu, R. Boukherroub, G. Manuel, *J. Organomet. Chem.* 1998, 554, 135–138; (e) G. Tagliavini, D. Marton, D. Furlani, *Tetrahedron* 1989, 45, 1187–1196; (f) L. Djakovitch, J. Eames, R.V.H. Jones, S. McIntyre, S. Warren, *Tetrahedron Lett.* 1995, 36, 1723–1726; (g) E. Adaligil, B.D. Davis, D.G. Hilmey, Y. Shen, J.M. Spruell, J.S. Brodbelt, K.N. Houk, L.A. Paquette, *J. Org. Chem.* 2007, 72, 6215–6223; (h) L.M. Grubb, B.P. Branchaud, *J. Org. Chem.* 1997, 62, 242–243; (i) K. Matsuo, T. Arase, S. Ishida, Y. Sakaguchi, *Heterocycles* 1996, 43, 1287–1300; (j) J. Moulines, A.-M. Lamidey, V. Desvergnes-Breuil, *Synth. Commun.* 2001, 31, 749–758; (k) T. Koga, Y. Aoki, T. Hirose, H. Nohira, *Tetrahedron Asymm.* 1998, 9, 3819–3824.

5. (a) B. Trost, I. Fleming (eds.), *Comprehensive Organic Synthesis: Selectivity, Strategy and Efficiency in Modern Organic Chemistry*, vol. 6, Pergamon, London, 1992, pp. 22–31; (b) K.P. Herlihy, D.S. Sagatys, *J. Chem. Res.* 1996, 2, 501–524; (c) J.P. Vozza, *J. Org. Chem.* 1959, 24, 720; (d) P. Kraft, Patent WO2008/148236, 2008, to Givaudan Sa; (e) A. Molnar, K. Felfoeldi, M. Bartok, *Tetrahedron* 1981, 37, 2149–2152; (f) K.K. Showa Denko, U.S. Patent 6372921, 2002; (g) N. Meyer, D. Seebach, *Chem. Berich.* 1980, 113, 1304–1319; (h) J.E. Holladay, J. Hu, X. Zhang, Y. Wang, U.S. Patent, 173654, 2007; (i) W.C. Brinegar, M. Wohlers, M.A. Hubbard, E.G. Zey, G. Kvakovszky, T.H. Shockley, R. Roesky, U. Dingerdissen, W. Kind, U.S. Patent 6639067, 2003; (j) J. Defaye, A. Gadelle, C. Pedersen, *Carbohydr. Res.* 1990, 205, 191–202; (k) A. Costa, J.M. Riego, *Synth. Commun.* 1987, 17, 1373–1376; (l) M.B. Smith, *March's Advanced Organic Chemistry*, Wiley Interscience, New York, 2001, pp. 479–480; (m) T. Zheng, R.S. Narayan, J.M. Schomaker, B. Borhan, *J. Am. Chem. Soc.* 2005, 127, 6946–6947; (n) M. Aquino, S. Cardani, G. Fronza, C. Fuganti, P.R. Fernandez, A. Tagliani, *Tetrahedron* 1991, 47, 7887–7896; (o) A.C. Spivey, A. Maddaford, T. Fekner, A.J. Redgrave, C.S. Frampton, *J. Chem. Soc. Perkin Trans. 1* 2000, 20, 3460–3468.

6. E.J. Alvarez-Manzaneda, R. Chaboun, E. Alvarez, E. Cabrera, R. Alvarez-Manzaneda, A. Haidour, J.M. Ramos, *Synlett* 2006, 12, 1829–1834.

7. (a) A.R. Katritzky, C.A. Ramsden, J.A. Joule, in *Handbook of Heterocyclic Chemistry*, Elsevier, Amsterdam, 2010; (b) A. Padwa, S. Bur, *Chem. Rev.* 2004, 104, 2401.

8. (a) K. Kindler, D. Matthies, *Chem. Ber.* 1962, 95, 1992; (b) T.M. Gadda, X.-Y. Yu, A. Miyazawa, *Tetrahedron* 2010, 66, 1249; (c) M. Haniti, S.A. Hamid, C. Liana Allen, G.W. Lamb, A.C. Maxwell, H.C. Maytum, A.J. Watson, J.M.J. Williams, *J. Am. Chem. Soc.* 2009, 131, 1766; (d) S.-I. Murahashi, K. Kondo, T. Hakata, *Tetrahedron Lett.*, 1982, 23, 229; (e) R. Grandel, W.M. Braje, A. Haupt, S.C. Turner, U. Lange, K. Dresher, L. Unger, D. Plata, Patent WO 2007/118899 A1, October 25, 2007, to Abbott Gmbh & Co. KG; (f) R.V. Radha, N. Srinivas, S.J. Kulkami, K.V. Ragavan, *Ind. J. Chem.* 1999, 38A, 286.

9. (a) N.R. Candeias, L.C. Branco, P.M.P. Gois, C.A.M. Afonso, A.F. Trindade, *Chem. Rev.* 2009, 109, 2703; (b) B. Ohtani, S. Tsuru, S. Nishimoto, T. Kagiya, *J. Org. Chem.* 1990, 55, 5551; (c) Y. Ju, R.S. Varma, *J. Org. Chem.* 2006, 71, 135.

10. (a) Asahi Kasei Chemicals Corporation, Patent WO2007/34669 A1, 2007; (b) S. Budavari (ed.), *The Merck Index*, 11th ed., Merck and Co., Rahway, NJ, 1989; (c) M. Wang, H. Wang, N. Zhao, W. Wei, Y. Sun, *Ind. Eng. Chem. Res.* 2007, 46, 2683.

11. (a) R.G. Pearson, *J. Am. Chem. Soc.* 1963, 85, 3533; (b) R.G. Pearson, J.J. Songstad, *J. Am. Chem. Soc.* 1967, 89, 1827; (c) R.G. Pearson, *Acc. Chem. Res.* 1990, 23, 1.

12. (a) P. Tundo, M. Selva *Acc. Chem. Res.* 2002, 35, 706; (b) P. Tundo, M. Selva, A. Perosa, S. Memoli *J. Org. Chem.* 2002, 67, 1071; (c) A.E. Rosamilia, F. Aricò, P. Tundo, *J. Org. Chem.* 2008, 73, 1559; (d) A. Rosamilia, F. Aricò, P. Tundo, *J. Phys. Chem. B* 2008, 112, 14525–14529 (e) P. Tundo, S. Memoli, D. Hérault, K. Hill, *Green Chem.* 2004,

6, 609; (f) P. Tundo, F. Aricò, A.E. Rosamilia, S. Memoli, *Green Chem.* 2008, 10, 1181; (g) P. Tundo, F. Aricò, G. Gauthier, L. Rossi, A.E. Rosamilia, H.S. Bevinakatti, R.L. Sievert, C.P. Newman, *ChemSusChem* 2010, 3, 566–570.

13. (a) F. Aricò, P. Tundo, A. Maranzana, G. Tonachini, *ChemSusChem* 2012, 5, 1578–1586; (b) F. Aricò, P. Tundo, *J. Chin. Chem. Soc.* 2012, 59, 11, 1375–1384; (c) P. Tundo, F. Aricò, G. Gauthier, A. Baldacci, *C. R. Chimie* 2011, 14, 652–655; H.S. Bevinakatti, C.P. Newman, S. Elwood, P. Tundo, F. Arico, Cyclic Ethers International Patent PCT/ GB2008/050567, January 22, 2009.

14. (a) P.N. Davey, L. Payne, L. Sidney, C. Tse, U.S. Patent 5821375, 1998; (b) D.H.R. Barton, S.I. Parekh, D.K. Taylor, C. Tse, U.S. Patent 5463089, 1994; (c) G. Knuebel, A. Bomhard, T. Markert, U.S. Patent 5811560, 1998.

15. G. Flèche, M. Huchette, *Starch—Stärke* 1986, 38, 26–30.

16. F. Aricò, U. Toniolo, P. Tundo, *Green Chem.* 2012, 14, 58–61.

17. C. Jiménez-González, C.S. Ponder, Q.B. Broxterman, J.B. Manley, *Org. Proc. Res. Dev.* 2011, 15, 912–917.

18. (a) J. Andraos, *Org. Process Res. Dev.* 2009, 13, 161–185; (b) J. Andraos, M. Sayed, *J. Chem. Educ.* 2007, 84, 1004–1011; (c) J. Andraos, in *The Algebra of Organic Synthesis: Green Metrics, Design Strategy, Route Selection, and Optimization*, CRC Press, Boca Raton, FL, 2012.

Acronym Index

Compound Index

Formula Index

Subject Index

Milton Keynes UK
Ingram Content Group UK Ltd.
UKHW040104071024
449327UK00019B/804